T0316813

Mergers and Acquisitions in the Global Brewing Industry

Corporate Finance and Governance

Herausgegeben von Dirk Schiereck

Band 10

PETER LANG

Frankfurt am Main · Berlin · Bern · Bruxelles · New York · Oxford · Wien

Ramit Mehta

Mergers and Acquisitions in the Global Brewing Industry

A Capital Market Perspective

PETER LANG
Internationaler Verlag der Wissenschaften

Bibliographic Information published by the Deutsche Nationalbibliothek
The Deutsche Nationalbibliothek lists this publication in the Deutsche Nationalbibliografie; detailed bibliographic data is available in the internet at http://dnb.d-nb.de.

Zugl.: Darmstadt, Techn. Univ., Diss., 2012

Cover Design:
Olaf Glöckler, Atelier Platen, Friedberg

D 17
ISSN 1869-537X
ISBN 978-3-631-62521-7

© Peter Lang GmbH
Internationaler Verlag der Wissenschaften
Frankfurt am Main 2012
All rights reserved.

www.peterlang.de

Preface

Nur wenige Branchen erscheinen so gut geeignet, die Wertgenerierung durch Unternehmenszusammenschlüsse im internationalen Kontext und unter wechselnden Wettbewerbsbedingungen zu untersuchen, wie die Brauereiindustrie. Das Wettbewerbsumfeld war für die Bierbrauer allein schon spannend, weil sich in den letzten zwanzig Jahren immer deutlicher ein enges Oligopol der ‚Big Four' abzeichnet, das die Biermärkte dominiert. Jede Akquisition durch einen der vier großen globalen Brauereikonzerne verschiebt das Machtverhältnis innerhalb dieser Gruppe, so dass Wettbewerbseffekte zu beobachten sein sollten.

Das Wettbewerbsumfeld war für die Brauer durch die Bedienung internationaler Märkte auch immer grenzüberschreitend. Unter solchen Konstellationen erscheinen internationale M&A-Transaktionen besonders attraktiv und sollten eigentlich zu positiven Reaktionen an den Kapitalmärkten führen. Aber passiert das auch? Untersuchungen zum Erfolg internationaler M&A-Transaktionen sind bis heute rar, der Kenntnisstand insbesondere zum Erfolg von Akquisitionen in der Brauereiindustrie ist noch begrenzter und die Übertragbarkeit der Evidenz zu den Erfahrungen aus anderen Branchen ist mehr als fraglich. Eine Analyse der kurz- und langfristigen Erfolgsdeterminanten von Unternehmenszusammenschlüssen in dieser Branche war also dringend erforderlich.

Die vorliegende Arbeit nimmt sich dieser bedeutenden Forschungslücke mit viel Liebe zum Detail und höchster Sorgfalt an. Ihr primäres Ziel war es, anhand von Marktdaten den Erfolg von internationalen M&A-Transaktionen in der Brauereiindustrie zu ermitteln und wesentliche Determinanten dieses Erfolgs zu bestimmen. So wird ein objektiver Kenntnisstand erreicht, auf dessen Basis sich fundierte Handlungsempfehlungen für die Industriepraxis ableiten lassen. Zudem finden sich damit auch für wirtschaftspolitische Entscheidungsträger wichtige Informationen zum Verständnis und zur Bewertung des sich weiter fortsetzenden Konzentrationsprozesses.

Herr Mehta kann die selbst gesetzten Ziele in seiner Dissertationsschrift bestens erfüllen. Die Arbeit enthält eine Fülle hoch interessanter Resultate und ist so geschrieben, dass es dem Leser große Freude machen wird, sie Seite für Seite bis zum Ende zu lesen. Ich wünsche der Arbeit eine weite Verbreitung.

Professor Dr. Dirk Schiereck

Acknowledgements

Writing the following doctoral thesis has been a challenging and rewarding experience. The progress of my research would not have been possible without the help of many people whom I would like to acknowledge here.

I would like to express my sincere gratitude to my doctoral supervisor Prof. Dr. Dirk Schiereck for his academic guidance and encouragement throughout the thesis. Besides providing valuable advice and instant feedback, he always managed to balance the progress of research with an extraordinarily constructive and collaborative working atmosphere. I am also thankful for the opportunity to hold lectures on corporate finance related topics and interact with students of TUD, which I thoroughly enjoyed. In addition, I am grateful to Prof. Dr. Ralf Elbert for agreeing to provide the second opinion on my thesis.

I would also like to thank my doctoral colleagues at TUD Malte Raudszus and Christoph Ettenhuber for engaging in constructive discussions on methodical and econometric issues.

Finally, I would like to thank my friends and family. In particular, I would like to thank my parents for their unconditional support and my brother Gaurav, who always encouraged me to pursue my goals. Without their continuous support I would not have been able to complete this thesis.

All remaining errors are mine alone.

Ramit Mehta

Content Overview

Table of Contents

List of Tables

List of Figures

List of Abbreviations

AB-Inbev	Anheuser-Busch Inbev
AmBev	Companhia de Bebidas das America
a.o.	among others
Avg.	Average
ARCH	Autoregressive Conditional Heteroscedasticity
big four	Four largest breweries worldwide: Anheuser-Busch Inbev, Heineken, SABMiller, and Carlsberg
Bil.	Billion
CAPM	Capital Asset Pricing Model
CAGR	Compound annual growth rate
DS	Datastream
DW	Durbin-Watson [statistic]
EBIT	Earnings before interest and tax
e.g.	exempli gratia (for example)
FF3F	Fama French 3 Factor Model
GARCH	Generalized Autoregressive Conditional Heteroscedasticity
HML	High minus low
i.e.	id est (that means)
KS	Kolmogorov-Smirnov [test]
M&A	Mergers and acquisitions
M-GARCH	Multivariate GARCH
Mil.	Million
OLS	Ordinary least squares [regression]
RoW	Rest of the World [countries]
S&N	Scottish & Newcastle
SIC	Standard industry classification
SDC	Security Data Company
SMB	Small minus big

Trans. Val.	Transaction value
TUD	Technische Universität Darmstadt
U.S.	United States [of America]
USD	United States [of America] Dollar
Wx	Wilcoxon [test]

List of Symbols

Greek symbols

A	Multivariate regression intercept
β	Systematic risk factor (beta)
γ	Coefficients to GARCH variables
Δ	Delta
δ	Coefficients to multivariate regressions variables
ρ	Correlation coefficient between an asset and the market
σ	Standard deviation (σ^2 denotes variance)
Φ	Long-term mean variance

Single letter Latin symbols

E	Expected [e.g., beta]
i	Individual stock or industry
m	Market return
N	Number of observations
t	Trading day
u	Daily return [of a stock or index]
x	Lower limit of event window
y	Upper limit of event window

Other symbols

AAB	Average abnormal beta
AAR	Average abnormal return
AB	Abnormal beta
AR	Abnormal return
$B4$	Big four [breweries]
BC	Bidder Champion

BHAR	Buy-and-Hold-Abnormal Return
BHR	Buy-and-Hold Return
CAB	Cumulative abnormal beta shift
CAR	Cumulative abnormal return
CAAB	Cumulative average abnormal beta shift
CAAR	Cumulative average abnormal return
CB	Cross-border M&A
D_A	Date
EM	Target in emerging market
HML	Fama and French (1992) value factor (high minus low)
MB	Multi Bidder
MM	Mature market [transactions]
PT	Public target
R_f	Risk free return
RS	Relative size (transaction volume / market value)
SC	Share component
SMB	Fama and French (1992) size factor (small minus big)
StD	Standard deviation
T_A	Target in Asia
T_{EE}	Target in Eastern Europe
T_{LA}	Target in Latin America
TV	Transaction volume
Yr	Year

1. Introduction

The global brewing industry has experienced a sharp increase in mergers and acquisitions (M&A) activity in recent years. Consolidation has and continues to be a major trend in the sector as multi-national breweries seek to expand their activities into new emerging markets. At the same time, declining mature markets (in particular Western Europe) and resulting pressure on profit margins have encouraged brewers to engage in M&A, in order to gain in scale and benefit from synergies.

While the brewing sector has historically shown tendencies towards industry consolidation, it is notable that the volume of M&A activity has considerably increased over the last few years. SABMiller's joint venture with Molson Coors in the autumn of 2007 marked the first of a series of landmark transactions that have significantly impacted and reshaped the structure of the global brewing sector. Other major transactions include the joint acquisition of Scottish & New-castle by Heineken and Carlsberg and InBev's USD 52 billion acquisition of Anheuser-Busch in 2008, which remains the largest transaction in the sector to date. Following these mega deals, many large brewers were left with a significant debt load and consequently, M&A activity slowed considerably in 2009. Momentum changed again in January 2010, when Heineken announced the acquisition of Mexican FEMSA Cerveza in a transaction valued at USD 7.6 billion. More recently, SABMiller completed the acquisition of Foster's, valuing the Australian brewer at USD 12.8 billion.

When compared to other industries the brewing sector stands out as a sector that has been particularly prone to industry consolidation and M&A. In contrast to many other industries the production, distribution and marketing of beer is characterized by a relatively high fixed cost base, resulting in high levels of operational leverage (Earlam et al., 2010). Therefore, many brewers have looked to gain in size in order to benefit from economies of scale. This is usually achieved by merging brewing sites or particular business units in order to realize significant cost savings and reduce potential excess capacities. Increased scale typically results in the ability to exercise a significant amount of market power (Schwankl, 2008), for instance vis-à-vis suppliers to negotiate more favorable terms or in negotiations with major retailers.

Thus, it is not surprising that the global beer market today is dominated by large national/multinational brewers rather than local, regional brewers. The four largest brewers Anheuser-Busch Inbev, Heineken, SABMiller and Carlsberg (the "big four") control about 50% of the global beer market and compete fiercely for market share. The unique competitive situation among the "big four"

1

is also reflected in strong rivalry for potential acquisition opportunities. Going forward research analysts are certain that consolidation will remain a major trend in the coming years and that the "big four" will further increase their market share (Earlam et al., 2010).

Even though the global brewing industry has gone through significant consolidation and seen a lot of M&A in recent years, empirical evidence remains scarce. Therefore, the objective of this thesis is to fill this research gap and provide academic M&A research on the capital market effects of M&A for managers of brewing companies, investors and financial researchers. In particular, this thesis aims to provide a comprehensive analysis of the shareholder wealth and risk effects to acquiring brewers as well as the rival effects among the "big four".

In order to achieve the research objective, the thesis comprises three individual empirical studies that address selected capital market relevant topics within the global brewing industry. In addition to the empirical studies a detailed industry analysis is provided upfront in order to establish a better understanding of the sector and its unique characteristics.

The first and second empirical studies focus on shareholder wealth effects of M&A to acquirers and rivals. The first study sheds light on the immediate announcement effects and the determinant drivers of short-term performance. The second study follows up on these results and analyzes whether the determined effects are sustainable in the long run. Finally, the third study shifts the focus to the capital market risk profile of acquirers and rivals, and specifically analyzes the immediate impact of M&A on systematic risk. Although different in focus and approach, all three studies contribute to the main challenge of developing a better understanding of global brewing M&A and its capital market implications.

In the following the reader is provided with additional details on the following chapters of the doctoral thesis:

Chapter 2, Research Foundations: Market Overview and Industry Analysis

Chapter 2 provides a brief overview of the development of the global beer market and analyzes the industry dynamics. The chapter intends to provide the research foundations for the following empirical studies (chapters 3-5) and create a general understanding for the industry specific rationale behind M&A and the competitive pressures among the leadings brewing groups. The analysis shows that many brewers have turned to cross-border M&A in emerging growth markets in an attempt to counter the decline in beer volumes in many mature markets. In general, M&A transactions have been encouraged by an extraordinary synergy potential in the sector, where increased size provides brewers with a significant competitive advantage.

Chapter 3, Study 1: Short-term Success of Mergers and Acquisitions

The first empirical study analyzes the short-term wealth effects of horizontal M&A transactions and investigates whether the global synergy and efficiency potential of the sector are reflected by capital markets in the form of abnormal stock price reactions to acquiring and close rival companies. Based on a sample of 69 takeover announcements between 1998 and 2010, the study documents significant positive announcement returns to acquiring brewers, which stands in contrast to previous cross-industrial studies and thus represents an outstanding attribute of the sector. Moreover, a number of determinant variables are detected that significantly impact short-term performance: the study finds a significant positive impact of domestic transactions as well as cross-border deals involving targets in emerging markets. Other identified value drivers include transaction size, acquirer size and the target's public status. Furthermore, significant negative rival effects are determined across the leading brewing groups, when missing out on a potential M&A opportunity. Overall, the findings confirm the unique competitive situation among the "big four" where capital markets value the successful search for M&A targets and at the same time punish rival companies for missing out.

Chapter 4, Study 2: Long-term Success of Mergers and Acquisitions

The second empirical study builds on the findings of chapter 3 and analyzes the long-term shareholder wealth effects of horizontal M&A on acquirers and rivals. The results show that in contrast to existing cross-industrial studies, which provide consistent evidence of value losses, acquirers in the brewing industry do not suffer from long-term value losses, following the engagement in M&A. Based on a sample of 66 transactions, the second study provides evidence of abnormal performance that is insignificant if not significantly positive, thus suggesting that capital markets also value the successful search for acquisition targets in the long run. In addition, the study documents a significant negative impact of cross-border transactions and cash transactions on acquirer performance. On the other hand, no significant abnormal long-term returns are found for transactions by the "big four", neither to successful acquirers nor to rivals missing out on an acquisition opportunity. Overall, the results suggest that the determined short-term shareholder value effects are not sustainable in the long run.

Chapter 5, Study 3: Consolidation and Changes in the Risk Profile of the Brewing Industry

While the previous studies are centered around the shareholder wealth effects of M&A, Study 3 shifts the focus of analysis on the risk implications of M&A. Specifically, the study sheds light on the question whether the M&A strategies employed by brewing companies are reflected by capital markets in the form of abnormal risk shifts to acquiring brewers and rival companies. Based on a sample of 75 takeover announcements between 1998 and 2010, the third study documents significant negative acquirer abnormal risk shifts for cross-border transactions, especially when targets are based in emerging markets. Furthermore, the study provides evidence of significant abnormal negative risk shifts to the "big four" when announcing an M&A transaction and significant positive abnormal risk shifts when missing out on a potential M&A transaction as rivals. Overall, the results underline the unique competitive situation in the sector and show that consolidation may result in contrary risk implications on different industry players.

Chapter 6, Conclusion

The final chapter consolidates the main findings from the three empirical studies and elaborates on key implications for managers, investors and financial researchers. The thesis concludes with ideas on potential fields for further academic research.

2. Research Foundations: Market Overview and Industry Analysis

2.1. Introduction

The shape of the global brewing industry has significantly changed over the last years. While globalization is a general trend in many industries, the globalization of the brewing industry has long been lagging behind. The opening of several key markets such as China, Russia, Eastern Europe and India coupled with the decline of beer volumes in many mature markets triggered a dramatic change in the market structures of the industry, as many brewers adopted aggressive internationalization strategies seeking new growth opportunities. As a consequence, the global brewing industry has seen strong consolidation and is now dominated by large multinational brewing groups. Anheuser-Busch Inbev, SABMiller, Heineken and Carlsberg, together the "big four", control approximately 50% of the global beer market. Industry analysts are certain that the consolidation process will continue and that the "big four" will further increase their market share (Earlam et al., 2010).

The aim of this paper is to analyze the global brewing industry in order to provide a sound understanding of the sector and its characteristics. In particular the fundamental market structures shall be reviewed to explain the consolidation process in the industry and the rationale behind M&A. Thus, with this paper we intend to provide the research foundations for the following empirical studies.

The remainder of the paper is structured as follows: Section 2.2 provides an overview of the global beer market including the definition of the market, market segmentation and market size. Section 2.3 provides an analysis of the industry dynamics, utilizing Michael Porter's five forces framework (Porter, 1980). In addition, the competitive landscape including the major global players is reviewed. Finally, Section 2.4 summarizes and concludes.

2.2. The Global Beer Market

2.2.1. Market Definition and Segmentation

For the purpose of this study the analyzed market shall be defined as the global market for lager beer, ales (dark beer), stout beer and non alcohol beer. A brewer in this study is defined as any company that generates the majority of its business through the production and sale of beer to retail chains (e.g., convenience

5

stores, supermarkets and liquor stores) and/or on-trade establishments (e.g., pubs, restaurants).

While beer is made from only four main ingredients (water, hops, yeast and grains), brewers can differentiate their offered products to a large extent. Beer can vary in both measurable (e.g., color, alcohol content) and hard to describe dimensions (e.g., flavour, aroma and palate). By altering the ingredients and the parameters in production process, there can be virtually unlimited types of beers. While there are several niche categories (e.g., stout beer), most beers fall into two main categories: lager beer and ales (dark beer). The primary distinction between these types of beers is the type of yeast used in the fermentation process. Ales generally use top-fermenting yeast, which means that the yeast floats on the surface and then settles to the bottom after a few days. Lagers, on the other hand, use bottom-fermenting yeast, which does not float to the surface before settling. Another distinction is the temperature at which the beer is fermented. While ales are fermented at higher temperatures (15-20°C), lagers are fermented much colder (c. 10°C) (Tremblay and Tremblay, 2009). In addition to the above mentioned classic alcoholic beers, many brewers today have added low and non alcoholic beers to their product portfolios.

Figure 2.1: Global share of beer categories by volume (2008)

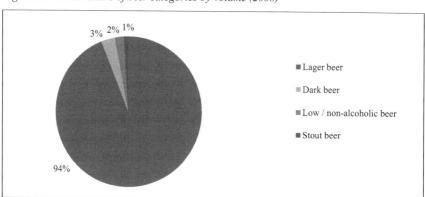

This figure presents the global market share of beer categories by volume in 2008. The data is sourced from an industry report from Euromonitor International (2009).

Figure 2.1 presents the global market shares of the most common beer categories as reported by Euromonitor International (2009). In the following we provide an overview of the reported beer categories:

Lager beer: Lager beer is by far the most important beer type, accounting for 94% of global beer sales by volume and 90% by value in 2008 (Euromonitor

International, 2009). While lager beer represents an entry level product in developing/emerging markets, in more developed markets it is generally consumed by younger and more fashion conscious consumers (Euromonitor International, 2009). A key development within the lager category has been the ongoing trend towards consumption of high-end, high quality lager (Diageo, 2010) particularly in well-established beer markets, where brewers are looking to compensate for lower volumes by selling higher-priced beer (Tremblay and Tremblay, 2009; Just-Drinks, 2007). The gain in market share is evidenced by the fact that between 2003 and 2008 the value growth of lager outperformed volume growth, clearly indicating a shift in demand towards premium and imported lager beer. On the other hand, the lager category has also seen the rise of lower-priced, economy lager beer, which has gained particular importance in emerging beer markets (Stout, 2007) but also in well-established beer markets such as the US (Greenberg and Kieley, 2009), Japan (Euromonitor International, 2011) or Germany (Rabobank Group, 2009). Despite the rise of premium and economy lager, globally standard lager remains the most important subcategory accounting for 50% of total lager beer in 2008 (Euromonitor International, 2009).

Dark beer: While dark beer is a traditional and popular beer category in a number of core markets including the UK, Germany and Belgium, it is considered a niche category on a global basis accounting for 3% of global beer volumes in 2008. In many markets dark beer suffers from an old-fashioned image and an ageing population (Euromonitor International, 2009). Nevertheless, recent years have shown a revival of dark beers in some of its major markets. According to Euromonitor International (2009) data, 52% of global dark beer volumes originate from Germany, the US and the UK. Germany, in recent years has seen a continuous growth in the consumption of wheat beer, a dark beer variety which has its origins in southern Germany, as wheat beer is becoming increasingly popular throughout many parts of the country, providing a viable alternative to the traditional lager variety Pilsener (Tholl, 2009). Similarly in the US dark beer has been the fastest growing beer category between 2003 and 2008, where it is increasingly regarded as a premium product (Euromonitor International, 2009). The positive trend can also be observed in the UK, where dark beer, in contrast to the general decline in industry beer volumes, is seeing an increase in consumption (Datamonitor, 2010).

Low-/non alcohol beer: While low-/non alcohol beer only accounted for less than 2% of global beer volumes in 2008, it has been the stand-out category in the sector both in terms of volume and value growth (Euromonitor International, 2009). The consumption of low-/non alcohol beer is in particular encouraged by the general decline in alcohol consumption in many developed markets (World Health Organization, 2011) as consumers are increasingly looking for "healthier" alternatives (Mintel, 2005). At the same time, low-/non alcohol beer has benefitted from stricter drink-driving regulations. A prominent example is Spain, where stricter drink-driving legislation was introduced in 2006. What started as a niche category, today accounts for approximately 20% of Spanish beer volumes and is by far the fastest growing beer category in the country, according to Euromonitor International (2009). In addition, the research firm expects significant category growth from Iran, Saudi Arabia, Nigeria and Egypt, which are among the major markets for non alcohol beer. In general, there has been a clear trend in these primarily Muslim countries to switch from standard soft drinks to non alcohol beer, which carries a fashionable, western image (Bates, 2009).

Stout beer: With a 1% share of global beer volumes in 2008, stout beer, a specialty dark beer, is clearly a niche category in the global beer market. The three biggest markets for stout beer Nigeria, Ireland and the UK accounted for 41% of global volumes in 2008. Stout volumes have been shrinking in traditional markets such as Ireland and the UK, as the premium priced stout beer is considered old-fashioned beer and consumers are switching to trendy import lagers (Euromonitor International, 2009). The Nigerian beer market, on the other hand, is a rare exception, offering the greatest potential for stout beer. Unlike in traditional markets, in Nigeria stout beer is mass-marketed in the low price, high volume segment. While global beer volumes for stouts have been declining, Nigeria saw stout volumes grow at a CAGR of 10.7% from 2003-2008 (Euromonitor International, 2009).

2.2.2. Market Size and Development

According to Euromonitor International data (2010), the global volume of beer consumed in 2009 amounted to 1.5bn barrels or 185bn liters. On average, the global beer market has remained relatively healthy between 2004 and 2009, with beer volumes growing at a compound annual growth rate of 4.2% (Greenberg, Fell, and Yordan, 2010). Table 2.1 presents the development of beer volumes in the 15 largest beer markets. Overall, growth prospects have been most attractive in Asia, Eastern Europe and Latin America, with China, Russia and Brazil accounting for 90% of volume growth alone. China, the world's largest market by size, stands out in particular with beer volumes growing at a 9% rate. At the same time, India saw the highest growth rate of 15.4% (Greenberg, Fell, and Yordan, 2010), however starting from a comparably low level. On the other hand, growth in more mature markets such as the USA, South Africa and Japan has significantly slowed down, with growth rates in the low single digits. Western European countries like Germany and the UK have even seen declining beer volumes as beer consumption per capita has been decreasing due to an ageing population and changes in consumption behavior (Euromonitor International, 2009). Overall, the top ten beer markets represent more than 70% of global beer volume.

Table 2.1: World's Largest Beer Markets by Volume (millions of barrels)

Rank	Country	2004	2009	Delta %	Rank: +/-
1	China	235.2	366.5	9.30%	0
2	USA	203.3	214.8	1.10%	0
3	Brazil	77.5	99.5	5.10%	1
4	Russia	76.3	93.0	4.00%	1
5	Germany	79.4	75.0	-1.10%	-2
6	Japan	58.5	60.0	0.50%	0
7	Mexico	46.8	55.3	3.40%	1
8	United Kingdom	51.7	42.6	-3.80%	-1
9	South Africa	27.7	29.7	1.40%	1
10	Poland	23.5	29.6	4.70%	1
11	Spain	27.9	27.5	-0.30%	-2
12	Venezuela	18.0	20.8	2.90%	1
13	Canada	18.9	20.7	1.80%	-1
14	Ukraine	14.7	20.5	6.90%	4
15	Romania	12.2	18.7	8.90%	8

This table shows the 15 largest beer markets by volume in 2009 and compares them to 2004 levels. Data source: Greenberg, Fell, and Yordan (2010) on basis of Euromonitor International data.

Table 2.2: World's Largest Beer Markets by Value (USD in millions)

Rank	Country	2004	2009	Delta %	Rank: +/-
1	USA	73,188	90,392	4.30%	0
2	Japan	50,796	57,387	2.50%	0
3	China	20,465	44,663	16.90%	2
4	Germany	31,239	32,995	1.00%	0
5	United Kingdom	37,239	25,852	-7.00%	-2
6	Brazil	9,426	22,814	19.30%	5
7	Russia	11,012	20,425	13.20%	1
8	Spain	16,498	18,891	2.80%	-2
9	Mexico	12,050	15,593	5.30%	-2
10	Canada	10,675	12,550	3.30%	-1
11	Venezuela	3,720	10,425	22.90%	11
12	France	9,567	9,499	-0.10%	-2
13	Italy	8,095	9,153	2.50%	2
14	Australia	9,095	8,936	-0.40%	-2
15	South Korea	8,986	8,136	-2.00%	-2

This table shows the 15 largest beer markets by value (USD) in 2009 and compares them to 2004 levels. Data source: Greenberg, Fell, and Yordan (2010) on basis of Euromonitor International data.

9

Table 2.2 presents the development of value growth in the ten largest beer markets. The results show that overall value growth kept up with volume growth, generating total revenues of USD 515 billion in 2009, which translates into a CAGR of 4.6% over the preceding five years. Similar to volume growth the main growth in value was derived from China, Brazil and Russia. Moreover, in contrast to volume growth, the US beer market also made up a significant part of the overall value gain with an estimated increase of USD 17 billion, second only to China's USD 23 billion increase. According to Greenberg, Fell, and Yordan (2010), research analysts at Deutsche Bank, profit pools in Western Europe, North America, Australia, and Japan remain quite large and may give rise to strategically attractive M&A opportunities. Other attractive profit pools include Mexico, which saw a significant rise in value growth despite a weak Mexican Peso. Within the top ten markets by value, only the UK declined by 7.0% as it was severely affected by declining volumes coupled with a significant decline in currency. Other well-established, mature markets such as Japan and Germany managed to grow in value despite declining volumes, which suggests that consumers shifted demand to higher priced, quality beers. Thus, in order to drive value growth it is becoming increasingly important to focus on strong branding and innovation strategies (Euromonitor International, 2009). Such strategies may also be relevant for Eastern European countries such as Poland, Ukraine or Romania, all of which hold a top 15 position in term of beer volumes but are not among the largest beer markets in terms of value.

Overall, the reported data shows that the beer markets across various countries are in different stages. Going forward it seems likely that international brewers will, on the one hand, look to focus their efforts on attractive growth markets such as Eastern Europe, Asia and Latin America. China and India in particular provide significant potential where beer consumption per capita is still relatively low. On the other hand, it can also be expected that, the well-established mature markets will be of interest, as some of them still remain highly interesting profit pools. Hence, the challenge of global brewers will be to balance their exposure between mature and emerging markets.

2.3. Industry Dynamics

2.3.1. Analysis of Porter's Five Forces

In the following the industry dynamics will be analyzed utilizing Michael Porter's five forces framework (Porter, 1980).

2.3.1.1. Intensity of Rivalry

According to Porter (1980), the concentration of an industry allows to draw conclusions on the intensity of rivalry in an industry. As pointed out in section 2.1, the global brewing industry has seen strong industry consolidation in recent years. A more complete picture is provided in Table 2.3. While in 2000 the top five brewers had a combined market share of 25.4%, they increased their share to 46.3% in 2009. A closer look at individual markets suggests even stronger consolidation: As Table 2.4 shows, the top three players in eight out of the ten largest beer markets by value hold a combined market share of at least 50%. Out of these eight markets, three (USA, Brazil and Mexico) have a clear market leader that alone controls at least 50% of the market. Thus, overall it seems fair to conclude that the global beer market can be regarded as fairly concentrated.

While industry consolidation generally suggests that rivalry in a sector is low (Porter, 1980), in case of the brewing industry, overall rivalry is relatively strong, in particular among the big brewing groups, which compete on price and quality aspects (Datamonitor, 2011). It is significantly boosted by the fact that consumers have a very wide range of products to choose from and thus switching costs are relatively low. While the rivalry in the premium sector is usually based on quality aspects, price-based rivalry can be observed in mass-marketed beers, which account for the majority of the large brewers' revenues (Datamonitor, 2011). Due to the necessity to operate large breweries and the resulting high fix cost base, brewers often look to benefit from economies of scale, which typically encourage rivalry, as high levels of production and potential excess capacity may lead to fights for market shares (Porter, 1980). Moreover, rivalry among brewers is further boosted by the development of the global beer market: On the one hand, it is strong in many mature beer markets, which are experiencing a significant decline in beer volumes (Shepherd, 1994). On the other hand, there is rivalry for new growth opportunities in emerging markets, as the large brewing groups compete for potential M&A targets in search for new growth opportunities. A prominent example includes the bidding war between Anheuser-Busch Inbev and SABMiller for the Chinese Harbin Brewery (Monaghan and Lim, 2004). Overall rivalry is assessed as moderate in the sector.

Table 2.3: Industry Concentration in the Global Beer Market

Year	Top five	Top ten	Others
2000	25.4%	37.4%	62.6%
2004	36.2%	48.0%	59.9%
2009	46.3%	59.9%	40.1%

This table presents the combined volume shares of the five and ten largest brewers in the global beer market. In addition the volume shares of the remaining brewers are provided. Euromonitor International (2010)

Table 2.4: Market Shares of Leading Brewers in ten Largest Beer Markets by Value (2009)

Country	Brewer	Share
1. USA	Anheuser-Busch Inbev	51%
	MillerCoors	29%
	Crown	6%
2. Japan	Asahi	38%
	Kirin	37%
	Suntory	12%
3. China	China Res (SAB Miller)	19%
	Tsingtao	13%
	Anheuser-Busch Inbev	11%
4. Germany	Oetker Group	15%
	Anheuser-Busch Inbev	10%
	Bitburger	9%
5. UK	Heineken	28%
	MolsonCoors	20%
	Anheuser-Busch Inbev	19%
6. Brazil	Anheuser-Busch Inbev	70%
	Schincariol	12%
	Heineken	9%
7. Russia	Carlsberg	38%
	Anheuser-Busch Inbev	18%
	Heineken	14%
8. Spain	Heineken	31%
	San Miguel	29%
	Damm	13%
9. Mexico	Modelo	55%
	Heineken	43%
	Anheuser-Busch Inbev	1%
10. Canada	Anheuser-Busch Inbev	43%
	MolsonCoors	41%
	Sapporo	4%

This table presents the market shares of the three largest brewers in the ten biggest beer markets by value. The data is based on a research report by Greenberg, Fell, and Yordan (2010), which utilizes Euromonitor International data for their calculations.

2.3.1.2. Bargaining Power of Buyers

Since the big brewing groups have over time gained in size, vertical integration can now be observed to some extent as brewers distribute their beer through self-owned on-trade establishments and distributorships. Nevertheless, for the most part of their sales, the big beer companies operate as non-vertically integrated businesses with retail and on-trade companies as their key buyers, which then sell the beer to the ultimate consumer. The degree of buyer power varies depending on the reliance of the buyer on beer sales to generate business. It is typically greater among big retail companies such as supermarket and hypermarket chains, than among on-trade establishments such as pubs, which would find it difficult to operate without beer. Especially within individual countries in Western Europe, supermarket chains are highly concentrated and can therefore negotiate strongly on price (Datamonitor, 2011). At the same time, buying power is increased as switching costs among buyers are low. While overall buyer power in the sector is assessed as strong, it is limited by a high degree of product differentiation and a wide range of established brands available that need to be offered by major buyers (Datamonitor, 2011).

2.3.1.3. Bargaining Power of Suppliers

As brewers have over time grown in size, some of them have started to use self-grown raw materials (e.g., hops) for beer production rather than buying them externally. Nevertheless, for the most part the big brewers still depend on external producers of malted grain, hops and bottles or barrels as key suppliers (see DeRise, 2009).

Supplier power is considerably limited by the fact that the required raw materials for beer production are commonly available. Hops are usually bought from independent hop growers which are numerous and usually small in size. Consequently, supplier power is decreased as the big brewing groups try to exert market power by bundling demand and negotiating favorable terms. On the other hand, supplier power is to some extent encouraged by the fact that some brewers may be dependent on the specialized, external supply of raw materials, in particular, as in the case of beer, the quality of the end product is considerably impacted by raw material quality. In addition, the market power of brewers is limited by the fact that some beer ingredients such as barley can be sold on a number of alternative markets including the animal food market as well as the market for spirits, where for example malted barley is used in the distillation process (Datamonitor, 2011). Nevertheless, overall supplier power in the sector is assessed as low.

2.3.1.4. Threat of Market Entry

Market entry into the beer sector on a large scale basis is made difficult by a number of entry barriers. In order to compete successfully with the established players, potential market entrants have to incur significant capital outlay on large production plants, as economies of scale provide existing high volume brewers with a material cost advantage. Economies of scale are of particular relevance as the brewing sector is characterized by high operating leverage (Earlam et al., 2010) allowing brewers to decrease unit costs considerably by increasing production volume. The size advantages are not limited to the production process as scale also helps brewers to secure raw material supplies and negotiate favorable terms with their suppliers, while market entrants, on the other hand, may struggle to establish reliable supplies. Furthermore, increased size provides brewers with a significant advantage with regard to their advertising and marketing efforts (Peles, 1971). At the same time, brand loyalty among consumers is regarded to be relatively high (Greenberg and Kieley, 2009). This makes it even more difficult to establish a new brand for a new market entrant. Potential market entrants may be further discouraged by government regulations, as in many countries the production of alcohol is taxed and may require a license (Datamonitor, 2011).

On the contrary, as the premium beer segment is gaining in importance, it has become possible to enter the market on a small-scale basis as a "microbrewery", where the initial high capital outlay can be recouped by pricing a strong margin onto the end product. A prominent example is the success of microbreweries in the US beer market, which shows that market entry into the premium segment is definitely possible (Carroll and Swaminathan, 2000).

Overall, the barriers to entry are high and thus the threat of potential market entrants is assessed as low.

2.3.1.5. Availability of Substitutes

Figure 2.2: Global Alcoholic Drinks Volume 2009 (in bn liters)

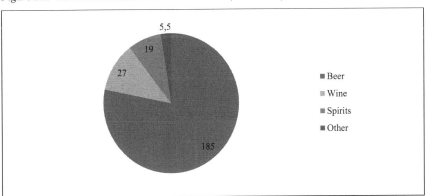

This figure presents the global market share of beer categories by volume in 2008. The data is sourced from an industry report from Euromonitor (2009).

With an estimated volume of 185bn liters in 2009, beer holds the majority of the market for alcoholic beverages with a clear lead over wine (27bn liters) and spirits (19bn liters) (Euromonitor, 2010). Nevertheless, over the past years, in particular some of the established beer markets (e.g., UK) have seen declining beer volumes as changing consumption patterns have led consumers to switch to other alcoholic drinks such as wine and spirits (Mintel, 2011), which act as the main substitutes for beer. At the same time, beer is seeing increased competition by non alcoholic beverages, as many consumers are becoming more health conscious (Mintel, 2005). The industry threat by substitutes is in particular encouraged by the fact that switching costs for consumers, retailers and on-trade businesses are practically non-existent. As many consumption decisions are a matter of personal taste, most brewers try to establish brand loyalty among consumers, by running large advertising campaigns (Spain, 2008). Over the years this strategy has so far proved successful as consumer loyalty among beer drinkers is regarded as relatively high (Greenberg and Kieley, 2009). Overall, the threat by substitutes is assessed as moderate.

2.3.1.6. Porter's Five Forces Summary

The industry analysis through Porter's five forces framework (Porter, 1980) has shown that the brewing sector can overall be regarded as an attractive sector to be active in. This finding is supported by the margin structure of the industry. As

15

presented in Figure 2.4, despite the difficult market environment in many markets, brewers have been able to expand their EBIT (Earnings before interest and tax) margins significantly from 2000 – 2009.

At the same time, the analysis has demonstrated the importance of scale advantages in the sector, which not only help the big brewers to compete among each other but also significantly strengthen their negotiation power with buyers and suppliers. Furthermore, as companies grow in size, threats by new market entrants become very difficult, as they need to incur significant capital expenditure to compete on the global market. Consequently, it is only logical that brewing companies have in the past looked at M&A in order to benefit from these size advantages and it seems very likely that they will continue to look at M&A in the future.

Figure 2.3: Porter's Five Forces Summary

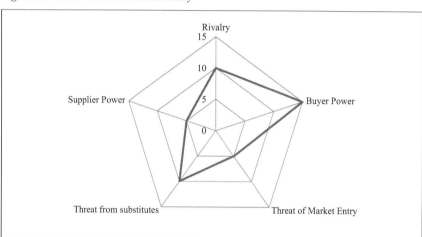

This figure provides a summary of the assessment of the dimensions of Porter's five forces framework 1980). "5", "10" and "15" indicate a low, moderate or high score, respectively.

16

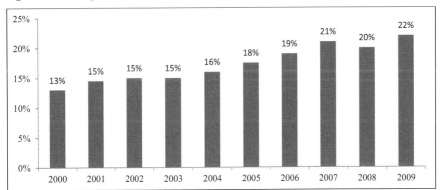

This figure presents the development of the average EBIT margins in the global brewing industry. Data source: DeRise (2009).

2.3.2. Key Industry Players

In order to provide a complete picture of the industry, the following section gives an overview of the leading companies in the sector.

The global brewing industry is dominated by four large brewing groups, namely Anheuser-Busch InBev, SABMiller, Heineken and Carlsberg (the "big four").

Figure 2.5: Beer Volumes of Leadings Brewers 2009 in Hectoliters

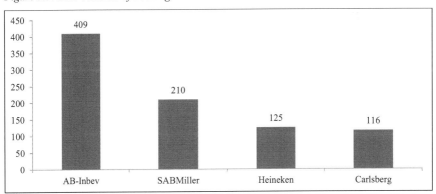

This figure presents the 2009 beer volumes of the "big four" (Anheuser-Busch Inbev, SABMiller, Heineken Carlsberg) in hectoliters. Data source: Annual Reports of "big four".

17

With estimated beer volumes of 409m hectoliters in 2009, Belgium-based Anheuser-Busch InBev (AB-InBev) is the clear industry leader. AB-InBev became the largest brewer in November 2008, when the Belgian brewing group InBev acquired US-based Anheuser-Busch for a total consideration of USD 52 billion. While the brewer has strong market positions across the globe, the majority of its earnings are generated from Latin America, where the company operates under the trade name of Companhia de Bebidas das America (AmBev). AB-InBev's brand portfolio comprises more than 300 brands including global flagship brands (e.g., Budweiser, Stella Artois and Beck's), multi-country brands (e.g., Leffe and Hoegaarden) as well as local domestic brands such (e.g., Bud Light, Skol and Quilmes). Furthermore, AB-InBev holds a 50% stake in Grupo Modelo, Mexico's leading brewer and owner of the global Corona brand and a 27% share in Chinese Tsingtao, whose same named beer brand, is among the best selling in China (Anheuser-Busch InBev, 2009).

SABMiller is the second largest brewer of the world with 210m hectoliters sold in 2009. The company's market presence ranges from established beer markets such as Europe, South Africa and North America to emerging beer markets such as Latin America, China and India. Nevertheless, the majority of the SABMiller's earnings are generated in Latin America and Europe. In July 2008 SABMiller joined forces with Molson Coors and entered into joint venture combining their operations in the US and Puerto Rico. More recently, SABMiller completed the acquisition of Australia's leading brewer Foster's. SABMiller's brand portfolio comprises leading premium brands such as Pilsner Urquell, Peroni Nastro Azzurro, Miller Genuine Draft as well a range of local brands including Aguila, Miller Lite and Snow (SABMiller, 2009).

With 125m hectoliters of beer sold in 2009, Heineken is the third largest brewer in the world. While Heineken has a wide international presence through a global network of breweries and distributors, the company generates the majority of its revenues in Europe (>75%), where it is the leading brewer in terms of volumes. In addition, Heineken has a strong position in Africa and the Middle East, which was its fastest growing segment in 2009. Following the recent acquisition of Mexican FEMSA Cerveza, Heineken has also strengthened its position in Latin America. Heineken's brand portfolio comprises more than 200 international premium, regional and local brands. Its principal brands include Amstel, Birra Moretti, Cruzcampo and Fosters (Heineken, 2009).

The fourth largest brewer Carlsberg reported sales of 116m hectoliters of beer in 2009. Unlike the other big brewers Carlsberg's business operations are focused entirely on Europe and Asia. The company generates more than 60% of its revenues in Northern and Western Europe. In addition, Carlsberg has strong market positions in Eastern Europe, in particular in Russia, where it is the leading brewer. Carlsberg's brand portfolio includes flagship brand Carlsberg Pilsener and strong regional brands such as Tuborg, Baltika and Kronenbourg 1664 (Carlsberg, 2009).

2.4. Summary and Outlook

The objective of this paper was to examine the global brewing industry in order to gain a better understanding of the sector and its particularities. For this purpose, a detailed assessment of the global brewing market and industry was provided. The key findings are summarized below:

First, the analysis of the markets segments has shown that, while there are many different beer categories, with virtually unlimited types of beers, the global beer market is dominated by lager beer, which accounts for over 90% of global beer volumes. Within the lager category there has been a strong trend towards premium lagers as well as economy lagers. While dark beer, the second largest beer category, is experiencing a revival in some countries, the fastest growing segment by far is low/non alcohol beer, which is encouraged by changes in consumption behavior and stricter drink-driving legislation.

Second, the analysis of global beer volumes has shown that volume growth has overall been strong, with most of the growth coming from emerging markets such as Asia, Eastern Europe or Latin America, where beer consumption per capita is still comparably low. On the other hand, mature beer markets such as Germany and the UK have experienced declining beer volumes due to market saturation and changes in consumption behavior. Nevertheless, in terms of value, many mature markets still remain interesting profit pools, as they offer superior margin profiles.

Third, the analysis of the industry dynamics has demonstrated the importance of scale in the sector, which provides brewers with a material competitive advantage. Thus, it is not surprising that the market is dominated by the "big four" who control about 50% of global beer volumes. At the same time, the analysis has shown that as a result of declining beer volumes in many mature markets there is strong rivalry among the "big four" for market shares and acquisition opportunities, primarily in emerging markets.

Having analyzed the brewing sector it seems clear why brewers have in the past turned to M&A strategies. M&A not only provide brewers with the opportunity to enter new growth markets in order to counter the decline in many mature markets, but also enable brewers to gain in size and strengthen their competitive position in an industry where size provides a material advantage. The following empirical studies specifically analyze M&A in the sector from a capital market perspective. More specifically, the studies investigate the wealth and risk effects to acquirers engaging in M&A transactions as well as the effects to rivals when missing out on acquisition opportunities.

3. Study 1: Short-term Success of Mergers and Acquisitions

3.1. Introduction

The analysis of shareholder wealth effects of M&A announcements is among the most extensively investigated research topics in empirical finance. While researchers unambiguously agree that overall the announcement of M&A transactions has positive value effects, Bradley, Desai, and Kim (1988) find that the short-term value creation is mostly attributed to the shareholders of target firms, which benefit from premiums paid by acquirers. On the other hand, evidence on wealth effects to acquiring firms is not conclusive. Even though Bruner (2002) concludes that overall acquirer returns average around zero, industry specific studies document mixed findings of negative abnormal performance, positive abnormal performance or abnormal performance that is not significantly different from zero. In addition to analyzing the wealth effects of the merging firms, there is also growing interest in the intra-industry effects of M&A announcements. The wealth implications for shareholders of rival companies are based on two opposing effects: positive information signaling regarding future takeover activity and negative competitive effects of the announced transaction. Existing empirical evidence suggests that the positive effects outweigh the negative effects as the majority of studies report rival company gains following M&A announcements (See Eckbo, 1983; Fee and Thomas, 2004; Shahrur, 2005; and Song and Walkling, 2000).

The global brewing industry has witnessed a sharp increase in M&A activity in recent years. Consolidation has and continues to be a major theme in the sector as multi-national brewing groups seek to diversify and expand their operations in new emerging markets. At the same time, declining beer volumes in many mature markets (in particular Western Europe) and resulting pressure on profit margins have encouraged brewers to pursue M&A strategies, in order to gain in scale and benefit from synergies. In contrast to many other industries the production, distribution and marketing of beer is characterized by a relatively large fixed cost base, resulting in high levels of operational leverage (Earlam et al., 2010) encouraging brewers to gain in size in order to benefit from economies of scale. In addition, increased size typically results in the ability to exercise market-power (Schwankl, 2008) for example vis-à-vis suppliers in order to negotiate more favorable terms or in negotiations with major customers (e.g.

retailers). Thus, it is not surprising that the global brewing industry today is dominated by large national/multinational brewing groups rather than local, regional breweries. The four largest brewers Anheuser-Busch Inbev, Heineken, SABMiller and Carlsberg (the "big four") control about 50% of global beer volumes (Jones, 2010). Despite the continued rise in market concentration, rivalry among the "big four" has remained fierce (Iwasaki, Seldon, and Tremblay, 2008). Going forward, industry experts anticipate that the consolidation process will continue (Fletcher, 2011). As sector debt levels are expected to decrease further, research analysts are certain that M&A activity will remain strong in the coming years, as brewers will continue their search for suitable M&A opportunities (Earlam et al., 2010).

In light of these specific industry characteristics and the ongoing consolidation process in the sector, the question arises whether the global synergy and efficiency potential of M&A transactions are reflected by capital markets in the form of abnormal stock price reactions to acquiring and close rival companies. Even though the global brewing industry has seen many M&A transactions in recent years empirical research remains scarce. Thus, the aim of this study is to fill this gap and provide empirical research for investors and managers of beer companies. In contrast to previous research, this study determines the short-term wealth effects of brewing companies based on a global dataset, utilizes a multi-factor model (Fama-French 3 Factor model) to determine statistically reliable indications of short-term performance, and investigates a comprehensive list of deal, acquirer and target characteristics for their impact on acquirer performance. In addition, the study assesses the unique competitive situation in the sector by specifically analyzing rival effects among the "big four".

The research objective of this study is twofold: first, we aim to update and extend previously published short-term wealth effects on acquirers in the brewing industry. The main focus lies in examining short-term return patterns in order to detect and categorize determinant variables. Second, we aim to provide empirical evidence with regard to M&A announcement effects among the "big four" as well as the wealth implications when missing out on transactions as rivals.

The remainder of this paper is structured as follows: Section 3.2 gives a brief overview of the relevant literature and outlines the derived hypotheses. Section 3.3 provides details on the applied methodology as well as the sample selection procedure. The following section 3.4 presents the empirical results and elaborates on the derived hypotheses. Finally, section 3.5 summarizes the findings and concludes.

3.2. Literature Review and Research Focus

3.2.1. Evidence on Wealth Effects around M&A

While there is general consensus among researchers that overall the announcement of mergers and acquisitions has positive value effects, Bradley, Desai, and Kim (1988) find that the short-term value creation is mostly attributed to the shareholders of target firms, which benefit from premiums paid by acquirers. On the other hand, studies that investigate acquirer returns provide evidence of short-term value losses: Based on a sample of 4,265 M&A transactions between 1973 and 1998, Andrade, Mitchell, and Stafford (2001) document insignificant negative returns to acquiring companies during a three-day event window surrounding the announcement date of the transaction. Similarly, Loughran and Vijh (1997) report short-term acquirer returns that are overall negative, or insignificant. After a comprehensive review of 44 individual studies, Bruner (2002) concludes that on average abnormal returns for acquirers are essentially zero.

In addition to the above-mentioned cross-industrial studies, a number of industry specific analyses seem to confirm the general results: Beitel, Schiereck, and Wahrenburg (2004) report negative abnormal acquirer returns for banks, while Akdogu (2009) and Berry (2000) document negative acquirer revaluations for the telecommunications and electric utilities industry, respectively. At the same time, researchers have identified certain "outlier" industries where acquirers are able to realize significant positive abnormal short-term returns. For instance, Mentz and Schiereck (2006) based on a sample of 201 M&A transactions in the automotive supply industry report significant abnormal returns to acquirers of +1.6% during a 10-day event window. The authors explain their finding with the extraordinary synergy potential in the industry perceived by capital markets. Similarly, Choi and Russel (2004) document positive abnormal returns to acquirers in the construction industry. Obviously, cross-industrial studies cover industry specific divergences, which result from the unique industry logic of value chains.

Besides investigating the impact of M&A on the merging firms empirical finance research also analyzes the implications of M&A on rival companies. Overall findings suggest positive as well as negative effects on rivals: On the one hand, rival companies may benefit from the M&A announcement due to positive signaling effects with regard to the attractiveness of the industry and potential future takeover activity (Eckbo, 1983; Song and Walkling, 2000). At the same time, a merger in the industry reduces the number of competing firms and thus increases the probability of collusion, which could potentially lead to greater monopoly rents to rival firms (Eckbo, 1983; Shahrur, 2005). Moreover, Snyder (1996) argues that rival firms may benefit from greater buyer power due to increased competition among suppliers, which in turn could lead to lower input prices. On the other hand, rival firms may be affected by negative competitive effects as a result of more intense competition in the industry from a new, more efficient combined firm (Eckbo, 1983). Overall, the documented positive effects outweigh the negative competitive effects: For instance, Eckbo (1985) reports positive announcement effects to rivals for horizontal transactions. Similar results are documented by Song and Walkling (2000) in a study including horizontal and non-horizontal transactions. More recent studies by Clougherty and Duso (2009), Fee and Thomas (2004) and Shahrur (2005) confirm these results.

As mentioned above empirical research on M&A in the brewing industry remains scarce and primarily focuses on the US brewing industry. The specific topics addressed in the studies focus on technological change in the sector (Kerkvliet et al., 1998), its tendencies towards concentration (Lynk, 1985; Adams, 2006), the determinants and motives for horizontal M&A (Tremblay and Tremblay, 1988), as well as competition in the industry (Horowitz and Horowitz, 1968). More recently, Ebneth and Theuvsen (2007) analyze the short-term value effects of M&A to acquirers using event study methodology. Based on a sample of 29 cross-border transactions involving European acquirers from 2000-2005, they find insignificant positive acquirer returns of 0.9% in the five-day event window surrounding the announcement date of the transactions.

3.2.2. Contribution to the Literature and Hypotheses

This study aims to contribute to and extend existing literature with regard to geographical scope and the methodology applied. First, we provide an analysis using a dataset, that in addition to cross-border transactions also includes domestic acquisitions and overall the merger wave of recent years. This database enables us to cover the complete M&A cycle where usually later transactions significantly differ from early ones. Second, we use a multi-factor (Fama French 3-factor model) model to determine and measure abnormal performance, which is then tested for significance using parametric and non-parametric statistical methods as well as multivariate regression analyses.

The main research interest is concentrated on the following aspects:

i) Overall Acquirer Wealth Effects:
As pointed out above, previous event studies focusing on single industries predominantly report negative acquirer returns. At the same time, certain indus-

24

tries have been identified as "outlier" industries reporting positive abnormal acquirer returns. Given the particularities of the brewing industry and its development over the last few years, we expect M&A in the sector to be a feasible measure to realize synergy and efficiency gains. Accordingly, we assume capital markets to reflect the industry-specific synergy potentials, resulting in positive short-term value effects to acquirers. Since our study is based on a larger dataset than used by Ebneth and Theuvsen (2007) and additionally considers global as well as domestic transactions, we assume significant positive abnormal returns to acquiring brewers.

ii) Determinants of Acquirer Performance:
While various studies focus on the impact of geography on short-term performance, the reported results are mixed: Several studies on cross-border M&A have found significant positive value gains to investors of acquiring firms around the announcement date (see e.g., Zhu and Malhotra, 2008; Goergen and Renneboog, 2004; and Morck and Yeung, 1992). On the other hand, there have also been studies documenting negative or insignificant gains to acquirers involved in cross-border M&A (see e.g., Datta and Puia, 1995; Eckbo and Thorburn, 2000). Negative acquirer returns are also reported for transactions involving targets in emerging market economies (Williams and Liao, 2008). In case of the brewing industry, Ebneth and Theuvsen (2007) find insignificant positive acquirer returns for cross-border transactions. However, their sample is restricted to 29 transactions. Due to the continuous decline in beer volumes in many mature markets, we regard cross-border M&A as a viable strategic option to diversify into international markets and thus expect significantly positive abnormal returns to acquirers. In particular, we assume acquirer returns to be positively impacted if the targets are based in emerging market economies and expect to find significant differences compared to domestic transactions.

The brewing industry has seen a significant increase in transaction volumes in recent years. Given the particular characteristics of the sector, increased company size provides brewers with a material competitive advantage. Analyzing technological change and economic efficiency in the U.S. brewing industry, Kerkvliet, et al. (1998) report substantial increases in economies of scale. Consequently, it can be argued that the acquisition of big targets will significantly contribute to the success of a transaction due to greater potential for economies of scale and revenue and/or cost synergies. On the other hand, the integration of larger targets may be more difficult than for small targets (Hawawini and Swary, 1991).

In addition to target size, the size of the acquirer may also influence the success of a transaction. Comprehensive studies by Asquith, Bruner, and Mullins (1983), Jarrell and Poulsen (1989) and Moeller, Schlingemann, and Stulz (2003) find a negative impact of acquirer size on acquirer returns. Moeller, Schlingemann, and Stulz (2003) argue that managers of larger companies are more likely to overestimate their own abilities. Due to the fact that larger companies may benefit from bigger cash reserves and might need to face less hurdles in the execution of M&A transactions, the authors argue that managers of larger companies are more likely to engage in M&A transactions that are not always beneficial to the company. Consequently, we expect to find significant differences in abnormal returns between large and small acquirers.

Over the last few years, the brewing industry has experienced a sharp increase in M&A activity and seen strong industry consolidation. As a consequence, the competitive landscape has materially changed as the "big four" today control more than 50% of global beer volumes. While the sector is expected to further consolidate in the future, it is becoming increasingly difficult to find suitable targets (Gibbs, Webb, and Dhillon, 2010). We assume the changes in market structures and concentration to have an impact on acquirer returns and expect to find significant differences in abnormal acquirer returns over time.

Several studies analyze the impact of the method of payment on acquirer returns. Myers and Majluf (1984) argue that bidders prefer to pay using stock when they believe that the market overvalues their shares and on the other hand prefer using cash, when they regard their stock as undervalued. Similarly, Martynova and Renneboog (2006) suggest that the means of payment used is an important signal of the quality of the target firm and its potential synergy value. They argue that a cash offer by the bidding company signals a willingness to pay off target shareholders in order to avoid sharing future cash flows and bear the sole risk of the combined firms. On the other hand, an all-equity offer signals the willingness to keep the target shareholders involved in the merged company and share its risk. The theoretical framework behind the mentioned signaling effects is supported by the results of different studies (see e.g., Brown and Ryngaert, 1991; Wansley, William, and Yang, 1983) that report a significantly negative market reaction following the announcement of equity offerings, in contrast to positive announcements returns in the case of cash offers. We expect to find similar results for the brewing industry.

The acquisition of privately held companies accounts for the majority of M&A transactions in the brewing industry. The general consensus among researchers is that bids for privately held companies generate higher bidder returns than bids for publicly held companies. Martynova and Renneboog (2006) argue that, in the case of privately held targets, bidders are likely to benefit from price discounts as compensation for buying a comparably illiquid stake. At the same time, they see advantages due to the fact that private companies usually have

26

fewer shareholders, which facilitates negotiations. These theoretical assumptions are confirmed in studies from Moeller, Schlingemann, and Stulz (2003) and Faccio, McConnell, and Stolin (2006) who report substantially higher announcement returns to acquirers for bids on privately held targets as opposed to bids on public firms. Consequently, we expect to find similar results for the brewing industry.

In a study focusing on short-term value effects of acquirers Haleblian and Finkelstein (1999) determine a positive relation between the number of completed transactions of an acquirer and the magnitude of the acquirer's abnormal return. They argue that with each completed transaction, acquirers gain experience in the integration of targets, which can be, leveraged in future transactions. In case of the brewing industry, which is dominated by large brewing groups that frequently engage in M&A activities, we expect to find greater returns for bidders with transaction experience.

iii) Announcement- and Rival Effects Among the "big four":
The global beer market is dominated by the "big four". Given the strong competition among them and due to the increased difficulty to find suitable targets, we expect to find positive acquirer returns if one of the "big four" announces an M&A transaction. At the same time, we expect the remaining three rival companies to be negatively impacted by the announcement as they are put into a disadvantageous competitive position by missing out on a potential M&A opportunity in a fairly concentrated market.

3.3. Data Selection and Research Methodology

3.3.1. Data Selection

The sample of mergers and acquisitions for the event study is drawn from the Securities Data Corporation (SDC)/ Thomson One Banker Deals database and the Merger Market M&A database. It includes all worldwide M&A events announced between January 1st, 1998, and September 1st, 2010. The total number of M&A deals is reduced to yield only those transactions meeting the following criteria:

i.   At the time of the transaction, acquirer and target companies both had active operations in the brewing industry.

ii. The acquiring company has been publicly listed for at least 250 days prior to the announcement of the transaction.

iii. The total transaction value accumulates to at least USD 50 million.

iv. The completion of the transaction leads to a change of control in the target; prior to the announcement of the transaction the bidder holds less than 50% in the target company, following the transaction the bidder obtains a controlling stake in the target company.

v. The transaction has been successfully completed.

In addition, the transactions were validated by a press research using the Factiva database as well as company websites in order to ensure that all transactions are horizontal and the announcement dates provided by the databases are correct. Moreover, acquirers with multiple transactions on the same day as well as acquirers with limited available trading data were removed from the dataset. The described selection criteria result in a final sample of 69 transactions. The frequency distribution of the transactions over time is provided in Table 3.1. While the number of transactions is spread fairly even over the years, the average transaction size varies strongly from USD 268 million to USD 11,973 million due to a number of high-profile transactions such as InBev's acquisition of Anheuser-Busch (USD 52 billion), Heineken and Carlsberg's takeover of S&N (USD 15 billion) and Heineken's recent acquisition of Femsa Cerveza (USD 5.7 billion). In terms of geography, more than 80% of the transactions involve acquirers that are based in Europe.

Table 3.1: Overview of the Transaction Sample: Descriptive Statistics

Year	Deals	(%)	Avg. Trans. Val. (USD mil.)	Trans. Val. (USD mil.)	Acquirer Region - Number of Deals			
					Europe	Americas	Asia	RoW
1998	4	5.8	268	1,071	4			
1999	4	5.8	495	1,981	3	1		
2000	6	8.7	514	3,086	5		1	
2001	4	5.8	599	2,396	3	1		
2002	6	8.7	1,526	9,157	5	1		
2003	6	8.7	582	3,489	6			
2004	9	13.0	875	7,873	6	2	1	
2005	8	11.6	785	6,281	8			
2006	6	8.7	332	1,991	6			
2007	5	7.2	405	2,025	1	2	1	1
2008	6	8.7	11,973	71,835	6			
2009	4	5.8	737	2,946	2		2	
2010	1	1.4	5,700	5,700	1			
Sum	**69**	**100.0**	**1,736.7**	**119,831.0**	**56**	**7**	**5**	**1**

This table provides the frequency distribution of the M&A transactions in the sample. It includes all successfully completed transactions between 1998 and 2010 where at the time of the transaction, acquirer and target companies both had active operations in the brewing industry, the acquiring company has been publicly listed for at least 250 days prior to transaction announcement, the total transaction value accumulated to at least USD 50 mil. and the bidder through the transaction acquired a controlling stake in the target company. Additionally, total and average transaction values (in USD mil.) and details on acquirer region are provided. Americas includes North and South America. RoW represents countries from the rest of the world.

The relevant daily stock prices, market capitalizations and local market indices for acquirers were downloaded from the Thomson Datastream database. Acquirer returns are calculated using the Datastream Total Return Index, which adjusts the closing share prices for dividend payments as well share issuances or repurchases. Moreover, global value and growth indices for large and small cap companies from data and index provider Russell serve as proxies for the Fama French model.

3.3.2. Research Methodology

In order to determine and analyze short-term announcement effects our study applies event study methodology. Event studies have a long history that goes back to the 1930s (see MacKinlay, 1997). Since then the methodology has become more and more sophisticated and found its application in empirical research on M&A. In particular, it has become a widely accepted tool to analyze the short-term value effects of M&A transactions. In our study, we assess short-term announcement returns using traditional event study methodology as for example described by Brown and Warner (1985) in connection with the "Fama-

French-3-Factor-model" (FF3F). The use of the multi-factor FF3F model enables us to more accurately detect and determine abnormal performance than with a single factor market model as used by Brown and Warner (1985). Equation 3.1 shows how the abnormal returns are derived:

$$E(R_i) = R_f + b_i[E(R_m) - R_f] + s_i E(SMB) + h_i E(HML) \qquad (3.1)$$

where $E(R_i)$ is the expected return on asset i, R_f is the return on the risk-free asset, $E(R_m)$ is the expected return on the market portfolio, $E(SMB)$ is the expected return on the mimicking portfolio for the "small minus big" size factor and $E(HML)$ is the expected return on the mimicking portfolio for the "high minus low" book-to-market factor.

Fama and French (1992), in probably one of the most influential papers in the area of asset pricing in the past decade, argue that the single factor Capital Asset Pricing Model of Sharpe (1964) and Lintner (1965) has little ability to explain the cross-sectional variation in equity returns. They find that two other factors related to fundamental variables, namely size and the ratio of book equity to market equity, have strong roles in explaining variation in cross-sectional returns. In our study, the multi-factor FF3F model is used to determine abnormal returns of acquiring and rival companies by regressing a time series of the companies' excess returns (return less risk-free rate) with the time series of market excess returns, the time series of the difference in returns of small and big companies (SMB), and the time series of differences in returns of companies with high and low (HML) market-to-book values (equation 3.2).

$$R_{i,t} - R_{f,t} = \alpha_i + b_i(R_{M,t} - R_{f,t}) + s_i SMB_t + h_i HML_t + e_{i,t} \qquad (3.2)$$

where $R_{i,t}$ is the realized return on asset i at time t, $R_{f,t}$ is the realized return on the risk-free asset at time t, $R_{M,t}$ is the realized return on the market portfolio at time t, SMB_t is the realized return on the mimicking portfolio for the size factor at time t and HML_t is the realized return on the mimicking portfolio for the book-to-market factor at time t.

The return of the market portfolio within the model usually refers to a market index that is associated with the particular security. In order to account for regional differences in industry returns and country-specific risk profiles our study determines local indices for each acquirer in the sample. For example, the DAX 30 index is used for German acquirer companies and the FTSE All Shares index is used for UK-based acquirer companies within the sample. Our study uses the 3-month U.S. T-bill rate as a proxy for the risk free rate.

The difference in returns of small and big companies as well as the difference in returns of companies with high and low market-to-book ratios is determined using global Frank Russell style portfolios as proposed by Faff (2003). The Russell style portfolios are utilized to create proxies for the Fama and French SMB and HML factors. Specifically, the style indices chosen are: (a) Global Russell large-cap Growth Index, (b) Global Russell large-cap Value Index, (c) Global Russell small-cap Growth Index, (d) Global Russell small-cap Value Index.

The Global Russell large-cap Growth Index (The Global Russell large-cap Value Index) measures the performance of the largest global companies with higher (lower) price-to-book ratios and higher (lower) forecasted growth values. Similarly, the Global Russell small-cap Growth Index (The Global Russell small-cap Value Index), measure the performance of global small-cap companies with higher (lower) price-to-book ratios and higher (lower) forecasted growth.

Having determined all relevant factors we finally estimate the acquirer return model by using a multivariate Ordinary Least Squares (OLS) regression over a 230 day estimation period starting at trading day t=-250 relative to the announcement date of the transaction. Finally, on the basis of these estimated FF3F Model parameters, we calculate the abnormal returns for all acquirer companies using different event windows.

To test for statistical significance of acquirers' abnormal returns this study employs three test statistics. First, we apply a simple parametric t-test. Second, we use a cross-sectional test as proposed by Boehmer, Musumeci, and Poulsen (1992). The cross-sectional test is commonly used in event study literature as it accounts for a potential event-induced increase in standard deviation. Third, since non-parametric test statistics can be more powerful than parametric t-statistics (see Serra, 2002; Barber and Lyon, 1996), we apply the Wilcoxon Signed Rank test to provide for a thorough statistical review.

3.4. Empirical Results

In the following, the empirical results of our analyses are presented. We start off reporting the results for the total acquirer sample. In order to determine potential drivers of abnormal performance, we then report the results of the univarate and multivariate analyses. Finally, we specifically present announcement and rival effects among the "big four".

3.4.1. Overall Acquirer Wealth Effects

Table 3.2 reports the short-term announcement effects of M&A transactions on the total sample of acquirers in the brewing industry. The results show that acquirers earn a significant 1.77% in the [-5;5] and 1.47% in the [-1;1] event windows surrounding the announcement date. As Figure 3.1 shows, the abnormal returns peak following the day of the announcement of the M&A transaction. In case of the [-5;5] event window the results are significant at the 5%-level for the t-statistics as well as the cross-sectional test, and significant at the 10%-level for the Wilcoxon test statistics. In case of the [-1;1] event window the results are significant at the 1%-level for the t-statistics and significant at the 5%-level for the cross-sectional test. These findings are in line with the expected results and confirm the exceptional characteristics of the brewing industry. The short-term value effects show that capital markets in fact value the extraordinary synergy potentials in the brewing sector.

Table 3.2: CAARs to Acquirers

Total Sample of Acquirers (N=69)

Event-Window	CAAR	t-Test			z-Test			Wx Test		
		t-value	p-value		z-value	p-value		z-value	p-value	
[-20; 0]	1.48%	1.66	0.10	*	2.04	0.05	**	1.38	0.17	
[-10; 0]	0.44%	0.60	0.55		0.67	0.51		0.40	0.69	
[-5; 0]	1.20%	1.96	0.05	**	1.79	0.08	*	1.43	0.15	
[-1; 0]	0.77%	1.52	0.13		1.22	0.22		0.53	0.60	
[0]	0.38%	0.93	0.36		0.70	0.49		0.22	0.82	
[0; +1]	1.08%	2.16	0.03	**	1.84	0.07	*	1.41	0.16	
[0; +5]	0.95%	1.56	0.12		1.43	0.16		1.30	0.19	
[0; +10]	0.52%	0.70	0.49		0.79	0.43		0.83	0.41	
[0; +20]	0.12%	0.12	0.90		0.42	0.68		0.11	0.91	
[-1; +1]	1.47%	2.51	0.01	***	2.23	0.03	**	1.41	0.16	
[-5; +5]	1.77%	2.37	0.02	**	2.26	0.03	**	1.81	0.07	*
[-10; +10]	0.57%	0.57	0.57		0.81	0.42		0.17	0.86	
[-20; +20]	1.13%	0.87	0.39		1.47	0.15		0.42	0.68	

This table shows the cumulative average abnormal returns (*CAARs*) to acquiring companies in mergers and acquisitions in the brewing industry. It contains all public acquirers whose trading data was available between 250 before and 20 days after transaction announcement. Statistical significance at the 10%, 5% and 1% level is denoted by *, ** and *** respectively. The statistical significance has been tested using a standard t-test, the cross-sectional test as proposed by Boehmer, Musumeci, and Poulsen (1992) (z-Test) and the Wilcoxon Signed Rank Test as described by Barber and Lyon (1996).

Figure 3.1: AARs of Acquirers Surrounding Transaction Announcement

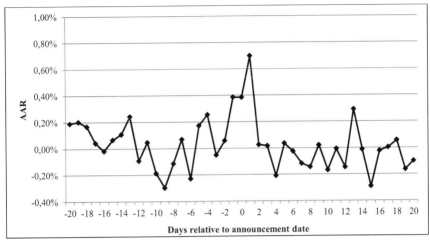

This figure provides daily abnormal average returns to acquiring companies of mergers and acquisitions in the brewing industry between 20 days before and 20 days after transaction announcement. It contains all public acquirers whose trading data was available between 250 before and 20 days after transaction announcement.

3.4.2. Determinants of Acquirer Performance

Geographical Scope: In order to analyze the impact of geographical diversification on acquirer returns we compare domestic transactions with cross-border transactions and emerging market transactions.

Table 3.3: ARs to Acquirers: Cross-Border vs. Domestic Transactions

	Event Window	[-1; +1]		[-5; +5]		[-10; +10]		[-20; +20]	
Cross-Border	CAAR	1.02%	*	0.86%		-0.70%		-0.36%	
	MEDIAN	0.54%		0.20%		0.11%		-2.41%	
	N	49		49		49		49	
Domestic	CAAR	2.56%		4.00%	**	3.69%	*	4.80%	**
	MEDIAN	-0.81%		1.79%	**	2.05%		4.77%	*
	N	20		20		20		20	
Difference	ΔCAAR	-1.54%		-3.14%	*	-4.39%	**	-5.16%	*
	ΔMEDIAN	1.35%		-1.59%		-1.94%		-7.18%	

This table shows the cumulative average abnormal returns (CAAR) to acquiring companies involved in cross-border and domestic mergers and acquisitions in the brewing industry. Statistical significance at the 10%, 5% and 1% level is denoted by *, ** and *** respectively. The statistical significance has been tested using the cross-sectional test as proposed by Boehmer, Musumeci, and Poulsen (1992) (z-Test) and the Wilcoxon Signed Rank Test as described by Barber and Lyon (1996).

Table 3.4: ARs to Acquirers: Emerging Market vs. Domestic Transactions

	Event Window	[-1; +1]	[-5; +5]		[-10; +10]		[-20; +20]	
Emerging Market	CAAR	1.05%	1.20%	*	1.94%	**	1.52%	
	MEDIAN	0.46%	1.46%		2.57%		0.09%	
	N	27	27		27		27	
Domestic	CAAR	2.56%	4.00%	**	3.69%	*	4.80%	**
	MEDIAN	-0.81%	1.79%	**	2.05%		4.77%	*
	N	20	20		20		20	
Difference	ΔCAAR	-1.51%	-2.80%		-1.75%		-3.27%	
	ΔMEDIAN	1.27%	-0.32%	*	0.52%		-4.68%	*

This table shows the cumulative average abnormal returns (CAAR) to acquiring companies involved in emerging market and domestic mergers and acquisitions in the brewing industry. Transactions are classified as emerging market transactions if the acquired target is based in Latin America, Asia (ex Japan) or Eastern Europe. Statistical significance at the 10%, 5% and 1% level is denoted by *, ** and *** respectively. The statistical significance has been tested using the cross-sectional test as proposed by Boehmer, Musumeci, and Poulsen (1992) (z-Test) and the Wilcoxon Signed Rank Test as described by Barber and Lyon (1996).

Tables 3.3 and 3.4 present the findings about the impact of geographical diversification on short-term acquirer performance. Overall, acquirers in the brewing industry show a preference for cross-border transactions and in particular emerging market transactions. In our total sample of 69 transactions, only 29% (20 transactions) are domestic/national transactions, while 71% (49 transactions) are cross-border transactions. Approximately 55% (27 transactions) of the cross-border transactions qualify as emerging market transactions and involve targets that are based in Latin America, Asia (ex Japan) or Eastern Europe.

On average domestic acquirers in almost every case show positive value effects upon the announcement of the transaction with *CAARs* and Medians of *CARs* ranging between -0.81% and 4.80%. Despite the small sample size of only 20 transactions, many of these returns are significant on the 5% and 10% level. On the other hand, acquirers in cross-border transactions show mixed effects with *CAARs* and Medians of *CARs* ranging between -2.41% and 1.02% across various event-windows. A comparison of means shows that *CAARs* for cross-border transactions are significantly lower for the [-5;5], [-10;10] and [-20;20]

35

event windows. Despite the relatively small sample size, these results indicate that capital markets seem to favor domestic over cross-border transactions. While these results stand in contrast to our predictions, they are in line with studies by Datta and Puia (1995) and Eckbo and Thornburn (2000). Nevertheless, the results will be challenged in the multivariate analysis due to the relative small amount of domestic transactions.

Table 3.4 compares domestic transactions with emerging market transactions. On average acquirers gain in all event windows upon the announcement of emerging market transactions with value gains ranging between 0.09% and 2.57%. Despite the small sample size of 27 transactions, acquirer *CAARs* gains are significant on the 5% and 10% level for the [-10;10] and [-5;5] event windows respectively. These results stand in contrast to the findings of Williams and Liao (2008), who report negative abnormal returns to acquirers in emerging market transactions in the banking industry. Again, due to the limited amount of transactions, these results will be challenged in the multivariate analysis.

Size of Transaction and Acquirer: In order to test for the incremental effect of transaction size on acquirer returns the sample was divided into two subsamples containing the 30 largest transactions and 30 smallest transactions by deal volume. The results are summarized in Table 3.5. On average, acquirers in small transactions yield positive *CAARs* between 0.94% and 2.47% across all event windows. In case of the [-5;5] event window, acquirers yield a positive 1.52% which is significant at the 5% level. On the other hand, acquirers in large transactions experience mixed value effects across various event windows with insignificant *CAARs* ranging from -2.67% to 1.14%. A comparison of means shows a significant underperformance of acquirer returns in case of large transactions for the [-10;10] and [-20;20] event windows. While these results stand in contrast to the expected results, they clearly serve as an indication and will be tested in the multivariate regression model.

Table 3.5: ARs to Acquirers: Top 30 vs. Bottom 30 Transactions by Size

	Event Window	[-1; +1]	[-5; +5]		[-10; +10]		[-20; +20]	
Top 30	CAAR	1.14%	0.88%		-2.23%		-2.67%	
	MEDIAN	0.59%	-1.24%		-1.78%	*	-4.41%	*
	N	30	30		30		30	
Bottom 30	CAAR	0.94%	1.52%	**	1.33%		2.47%	*
	MEDIAN	-0.66%	1.29%	**	0.94%		0.23%	
	N	30	30		30		30	
Difference	ΔCAAR	0.19%	-0.64%		-3.56%	*	-5.14%	*
	ΔMEDIAN	1.25%	-2.53%		-2.71%	**	-4.64%	***

This table shows the cumulative average abnormal returns (CAAR) to acquiring companies for the top 30 and bottom 30 transactions by transaction volume in the sample. Statistical significance at the 10%, 5% and 1% level is denoted by *, ** and *** respectively. The statistical significance has been tested using the cross-sectional test as proposed by Boehmer, Musumeci, and Poulsen (1992) (z-Test) and the Wilcoxon Signed Rank Test as described by Barber and Lyon (1996).

In order to analyze the impact of acquirer size, the transactions in the sample were sorted by relative deal size (transaction volume/acquirer's market capitalization) and divided into two subsamples containing the 30 largest and 30 smallest transactions. Table 3.6 presents the findings. We find insignificant positive *CAARs* to acquirers in the case of relatively large targets ranging from 0.70% to 0.95%. In the case of small transactions from the acquirer's perspective, the *CAARs* show greater variance ranging from 0.61% to 3.33% and are significant at the 10% level for the [-1;1] and [-5;5] event windows. While the differences are not significant, the results provide an indication of higher *CAARs* for small acquirers or the acquisition of relatively large targets as suggested by Asquith, Bruner, and Mullins (1983), Moeller, Schlingemann, and Stulz (2003) and Jarrell and Poulsen (1989), they will be challenged in the multivariate analysis.

Table 3.6: ARs to Acquirers: Top 30 vs. Bottom 30 Transactions by Relative Transaction Size

	Event Window	[-1; +1]		[-5; +5]		[-10; +10]	[-20; +20]
Top 30	CAAR	2.17%	*	3.33%	*	0.61%	1.41%
	MEDIAN	0.30%	*	2.05%		0.18%	-3.50%
	N	30		30		30	30
Bottom 30	CAAR	0.95%		0.70%		0.73%	0.80%
	MEDIAN	0.41%		0.61%		1.17%	0.23%
	N	30		30		30	30
Difference	ΔCAAR	1.22%		2.63%		-0.12%	0.61%
	ΔMEDIAN	-0.11%		1.43%	**	-0.98%	-3.73%

This table shows the cumulative average abnormal returns (CAAR) to acquiring companies for the top 30 and bottom 30 transactions by relative transaction size in the sample. Statistical significance at the 10%, 5% and 1% level is denoted by *, ** and *** respectively. The statistical significance has been tested using the cross-sectional test as proposed by Boehmer, Musumeci, and Poulsen (1992) (z-Test) and the Wilcoxon Signed Rank Test as described by Barber and Lyon (1996).

Time Period: In section 3.2 we argued that the change in market concentration and in particular the increased difficulty in finding suitable targets may have an impact on acquirer returns. Table 3.7 presents our findings comparing transactions between 1998 and 2003 and 2004 and 2010. While transactions announced between 1998 and 2003 yield insignificant positive and negative value effects, transactions between 2004 and 2010 on average yield positive value effects across all event windows. Moreover, with the exception of event window [-10;10] all of the *CAARs* reported are significant on the 5% or 10% level. A mean comparison, though not statistically significant, reveals higher returns to acquirers between 2004 and 2010 across all event windows. Despite the lack of statistical significance, these results serve as an indication and will be tested in the regression analysis.

Table 3.7: ARs to Acquirers:"1998-2003" vs. "2004-2010"

	Event Window	[-1; +1]		[-5; +5]		[-10; +10]		[-20; +20]	
1998-2003	CAAR	1.21%		1.64%		-0.62%		-0.38%	
	MEDIAN	-0.34%		0.66%		-0.17%		-3.35%	
	N	30		30		30		30	
2004-2010	CAAR	1.66%	**	1.87%	*	1.49%		2.30%	*
	MEDIAN	0.46%		1.19%	*	1.29%		0.09%	
	N	39		39		39		39	
Difference	ΔCAAR	-0.45%		-0.23%		-2.12%		-2.68%	
	ΔMEDIAN	-0.80%		-0.53%		-1.45%		-3.44%	

This table shows the cumulative average abnormal returns (CAAR) to acquiring companies for mergers and acquisitions in the brewing industry between 1998-2003 and 2004-2010. Statistical significance at the 10%, 5% and 1% level is denoted by *, ** and *** respectively. The statistical significance has been tested using the cross-sectional test as proposed by Boehmer, Musumeci, and Poulsen (1992) (z-Test) and the Wilcoxon Signed Rank Test as described by Barber and Lyon (1996).

Public Status of Target: Table 3.8 presents the results of acquirers' abnormal returns comparing transactions involving public and private targets. The acquisition of public targets yields mixed results with acquirers' *CAARs* ranging between -1.01% and 1.31% all of which are statistically insignificant. On the other hand, acquirers of private targets experience positive value effects with significant *CAARs* ranging between 1.58% and 2.49%. Moreover, the *CAAR* of 2.49% for the [-5;5] event window is statistically significant on the 1% level. While the mean comparison shows that across all analyzed event-windows the acquisition of private targets leads to higher acquirer returns, the differences are not statistically significant. Nonetheless, these results are in line with studies by Moeller, Schlingemann, and Stulz (2003) and Faccio, McConnell, and Stolin (2006).

Table 3.8: ARs to Acquirers: Public Targets vs. Private Targets

	Event Window	[-1; +1]		[-5; +5]		[-10; +10]	[-20; +20]	
Public	CAAR	1.31%		0.77%		-1.01%	0.02%	
	MEDIAN	0.06%		-1.56%		0.15%	-0.95%	
	N	29		29		29	29	
Private	CAAR	1.58%	*	2.49%	***	1.72%	1.94%	*
	MEDIAN	0.45%		1.61%	*	0.94%	-0.65%	
	N	40		40		40	40	
Difference	ΔCAAR	-0.28%		-1.72%		-2.73%	-1.92%	
	ΔMEDIAN	-0.39%		-3.17%		-0.79%	-0.30%	

This table shows the cumulative average abnormal returns (CAAR) to acquiring companies for mergers and acquisitions in the brewing industry for publicly-listed and private target companies. Statistical significance at the 10%, 5% and 1% level is denoted by *, ** and *** respectively. The statistical significance has been tested using the cross-sectional test as proposed by Boehmer, Musumeci, and Poulsen (1992) (z-Test) and the Wilcox-on Signed Rank Test as described by Barber and Lyon (1996).

Type of Consideration: Table 3.9 compares the results of acquirer's abnormal returns of cash only transactions with transactions that use share-based or hybrid forms of consideration. Overall, brewers show a clear preference for cash only transactions. On average, acquirers paying solely with cash experience a positive CAAR of 1.37% for the [-1;1] event window, which is significant on the 5% level. The limited amount of share deals does not allow for a viable comparison and will hence be addressed in the multivariate analysis.

Table 3.9: ARs to Acquirers: Cash Only Transactions vs. Share Transactions

	Event Window	[-1; +1]		[-5; +5]		[-10; +10]	[-20; +20]
Cash Only	CAAR	1.37%	**	1.08%		0.32%	1.03%
	MEDIAN	0.21%		0.88%		0.47%	-0.83%
	N	52		52		52	52
Share Deals	CAAR	2.09%		4.88%		2.31%	3.71%
	MEDIAN	0.54%		2.23%		2.56%	-0.95%
	N	11		11		11	11
Difference	ΔCAAR	-0.72%		-3.80%	*	-1.99%	-2.68%
	ΔMEDIAN	-0.33%		-1.35%		-2.09%	0.12%

This table shows the cumulative average abnormal returns (CAAR) to acquiring companies for mergers and acquisitions in the brewing industry for cash-only and share-based transactions. Statistical significance at the 10%, 5% and 1% level is denoted by *, ** and *** respectively. The statistical significance has been tested using the cross-sectional test as proposed by Boehmer, Musumeci, and Poulsen (1992) (z-Test) and the Wilcoxon Signed Rank Test as described by Barber and Lyon (1996).

Transaction Experience: Tables 3.10 and 3.11 present the results of acquirers' abnormal returns based on transaction experience. Overall, 13 of the total of 69 transactions involve acquirers that have only engaged in one transaction in the sample period (Single-Bidder) while 56 transactions involve acquirers that have engaged in at least one transaction in the sample period (Multi-Bidder). 36 transactions involve bidders that have engaged in more than five transactions in the sample period (Bidder-Champion). While we find no significant returns for Multi-Bidder transactions, we find a positive *CAAR* of 1.25% for Bidder-Champion transactions in the [-5;5] event window, which is significant at the 5% level. Due to the limited amount of Single-Bidder transactions, we shall provide a viable comparison in the regression analysis.

Table 3.10: ARs to Acquirers: Single-Bidder Transactions vs. Multi-Bidder Transactions

	Event Window	[-1; +1]		[-5; +5]		[-10; +10]	[-20; +20]
Single-Bidder	CAAR	3.38%	**	4.34%	*	1.53%	2.22%
	MEDIAN	1.88%		2.27%		-3.53%	-1.23%
	N	13		13		13	13
Multi-Bidder	CAAR	1.02%		1.17%		0.35%	0.88%
	MEDIAN	0.21%		0.88%		1.17%	-0.60%
	N	56		56		56	56
Difference	ΔCAAR	2.36%		3.17%	*	1.18%	1.35%
	ΔMEDIAN	1.67%		1.39%		-4.70%	-0.63%

This table shows the cumulative average abnormal returns (CAAR) acquiring companies for mergers and acquisitions in the brewing industry comparing single-bidders i.e., acquirers with only one announced transaction in the sample with multi-bidders i.e. acquirers with more than one announced transactions in the sample. Statistical significance at the 10%, 5% and 1% level is denoted by *, ** and *** respectively. The statistical significance has been tested using the cross-sectional test as proposed by Boehmer, Musumeci, and Poulsen (1992) (z-Test) and the Wilcoxon Signed Rank Test as described by Barber and Lyon (1996).

Table 3.11: ARs to Acquirers: Single-Bidder Transactions vs. Bidder-Champion Transactions

	Event Window	[-1; +1]		[-5; +5]		[-10; +10]	[-20; +20]
Single-Bidder	CAAR	3.38%	**	4.34%	*	1.53%	2.22%
	MEDIAN	1.88%		2.27%		-3.53%	-1.23%
	N	13		13		13	13
Bidder Champions	CAAR	1.25%	**	0.79%		-0.81%	-1.63%
	MEDIAN	0.50%		0.23%		0.04%	-3.89%
	N	36		36		36	36
Difference	ΔCAAR	2.14%		3.55%	*	2.34%	3.86%
	ΔMEDIAN	1.38%		2.04%		-3.57%	2.66%

This table shows the cumulative average abnormal returns (CAAR) acquiring companies for mergers and acquisitions in the brewing industry comparing single-bidders i.e., acquirers with only one announced transaction in the sample with bidder-champions i.e., acquirers with more than five announced transactions in the sample. Statistical significance at the 10%, 5% and 1% level is denoted by *, ** and *** respectively. The statistical significance has been tested using the cross-sectional test as proposed by Boehmer, Musumeci, and Poulsen (1992) (z-Test) and the Wilcoxon Signed Rank Test as described by Barber and Lyon (1996).

Multivariate Regression Analysis:

In order to provide a complete picture of the influential factors and to gain further insights into potential dependencies, a cross-sectional regression is performed on the cumulative abnormal returns to acquirers as presented in equation 3.3. In total, 11 variables are included in the regression model to represent the parameters, which have been individually analyzed in the univariate subsample analysis. In the following, the respective parameter values will be specified in detail.

$$CAR = \alpha_0 + \delta_1 * CB + \delta_2 * T_{LA} + \delta_3 * T_{EE} + \delta_4 * T_A + \delta_5 * TV + \delta_6 * RS$$

$$+ \delta_7 * D_A + \delta_8 * SC + \delta_9 * PT + \delta_{10} * MB + \delta_{11} * BC$$

(3.3)

Geographical Scope: The results presented in the univariate analysis provide a first indication that domestic transactions might have a positive impact on short-term acquirer performance when compared to cross-border transactions. At the same time, the results suggested a positive wealth effect if targets were based in emerging market economies (Latin America, Eastern Europe and Asia (ex Japan)). Both effects are included in the regression model using the dummy variables CB, T_{LA}, T_{EE}, T_A.

Size of Transaction and Acquirer: The results presented in the univariate subsample showed significant positive returns for the bottom 30 transactions by transaction value as well as the top 30 transactions by relative transaction value, indicating a preference for small transactions and small-sized acquirers. In order to verify these results, transaction value (TV) and relative transaction value (RS) are included as variables in the regression model.

Time Period: As pointed out above the structures of the global beer market have materially changed in recent years. Consequently, we expected these changes to have an impact on acquirer performance. In the univariate subsample analysis we found significant positive returns for transactions between 2004 and 2010, we did not find any significant abnormal performance between 1998 and 2003. In order to test for a supposed relation, we include a dummy variable for time period 1998 – 2003 (D_A).

Public Status of Target: The univariate results showed highly significant positive returns for the acquisition of private targets. On the other hand, no significant abnormal performance was found for public targets. We test these results by including a dummy variable for public targets in the regression model (PT).

Type of Consideration: The results provided in the univariate analysis showed significant positive returns for cash transactions for the [-1;1] event windows surrounding the announcement date. On the other hand, transactions with share-based consideration showed even greater abnormal returns, albeit being statistically insignificant. In order to test for a supposed relation, we include a dummy variable considering share-based consideration (SC).

Transaction Experience: The subsample analysis showed significant positive abnormal returns for bidder champions i.e., acquirers with at least five announced transactions in the sample. On the other hand, single bidders i.e., acquirers with only one announced transaction in the sample on average experienced even higher abnormal returns. In order to test for a potential relationship between transaction experience and acquirer return, we include two dummy variables reflecting Multi-Bidder (MB) and Bidder-Champion (BC) transactions.

Table 3.12: Multivariate Regression Analysis

	[-1; +1]		[-5; +5]		[-10; +10]	
Variable	Coefficient	t-value	Coefficient	t-value	Coefficient	t-value
Constant	0.047 ***	2.740	0.064 ***	2.994	0.047	1.623
CB	-0.014	-0.744	-0.021	-0.955	-0.050 *	-1.657
T_{LA}	0.037 **	1.982	0.025	1.104	0.044	1.408
T_{EE}	-0.019	-1.036	-0.008	-0.378	-0.014	-0.482
T_A	-0.005	-0.259	-0.010	-0.432	0.033	1.037
TV	0.000 *	1.915	0.000 **	2.074	0.000	0.829
RS	-0.018 *	-1.660	-0.029 **	-2.153	-0.040 **	-2.174
D_A	0.002	0.127	-0.003	-0.178	-0.012	-0.602
SC	0.015	0.873	0.059 ***	2.863	0.044	1.609
PT	-0.016	-1.147	-0.036 **	-2.167	-0.036	-1.603
MB	-0.029	-1.592	-0.022	-0.985	0.021	0.697
BC	0.009	0.550	-0.006	-0.278	-0.015	-0.529
R-squared	0.221		0.278		0.267	
Adjusted R-squared	0.071		0.139		0.126	
Durbin-Watson	1.990		2.379		2.530	
F-statistic	1.469		1.997 **		1.890 *	
p (F-stat)	0.169		0.046		0.060	

CAARs were derived for a sample of 69 transactions in the brewing industry between 1998 and 2010. For a detailed description of the variables and the underlying equation, see section 3.4.2. The Durbin-Watson statistics were estimated to test for autocorrelation of the residuals. Proximity of the value to "2" is regarded as an indication of no autocorrelation between residuals. Statistical significance at the 10%, 5% and 1% level is denoted by *, ** and *** respectively.

Table 3.12 presents the results of the complete regression models on the CAARs for the [-1;1], [-5;5] and [-10;10] event windows. The [-5;5] and [-10;10] models are significant on the 5% and 10% level respectively. Explanatory power is remarkably high with adjusted R-squared ranging between 12% and 14%. Autocorrelation issues can be ruled out due to high Durbin-Watson-Statistics in both cases. The overall abnormal short-term performance as represented by the constant yields a positive 4.7% for the [-1;1] and 6.4% for the [-5;5] event windows and are both highly significant at the 1% level. Overall, these findings correspond to the positive announcement effects determined in the univariate analysis, clearly confirming our expectations.

In addition, the regression models complement the univariate subsample analysis enabling the detection of a number of different value drivers of short-term performance. First of all, transaction value and acquirer size are determined to have a significant positive impact on short-term performance. These results stand in contrast to the findings of the univariate subsample analysis. The regression model confirms both target and acquirer size to be positively related to acquirer performance. With regard to target size, this relation can be confirmed for two of the regression models. On the other hand, the negative impact of relative deal size (positive impact of acquirer size) is confirmed across all three regression models. Overall, these findings provide clear evidence on the importance of size and its advantages in the brewing sector.

While not significant for all models, the multivariate analysis confirms the negative impact of cross-border transactions on acquirer returns. With regard to emerging market transactions, regression coefficients across all models are in most cases higher than those for all cross-border transactions. In the case of Latin America, the regression model for the [-1;1] event window even yields a positive regression coefficient of 3.7% which is significant at the 5% level.

The subsample analysis on the impact of time on acquirer returns suggested an increase in abnormal acquirer returns over time. The multivariate regression does not show any significant impact of the announcement date on acquirer returns.

With regard to the public status of the target, the subsample analysis suggested acquirer returns to be positively impacted if the acquired target was not publically listed. The regression models confirm these results with negative coefficients across all three event windows. Moreover, a negative relation is determined for the [-5;5] event window which is significant at the 5% level.

Despite the small set of non-cash transactions, the subsample analysis suggested acquirer returns to be positively affected if the transaction includes share-based compensation. The regression models confirm these results with positive coefficients throughout all three models. Nonetheless, only one of the coefficients is significantly positive and hence we do not believe there is enough evidence to confirm a potential dependency.

With regard to transaction experience, the univariate results suggested higher announcement returns to Single-Bidders when compared to Multi-Bidders. At the same time, we found significant positive returns to Bidder-Champions. The multivariate analysis does not provide any additional insight as to a potential dependency between acquirer return and transaction experience. Given the lack of additional evidence, we cannot confirm an impact of transaction experience on acquirer performance.

3.4.3. Announcement and Rival Effects among the "big four"

While the analyses presented above covered all publicly listed acquirers, in the following we give particular attention to announcement effects of the "big four" and rival effects among them. Table 3.13 presents the short-term announcement effects of the "big four". Overall, we find positive announcement returns of 1.25% for the [-1;1] event window. For the t and z statistics, these results are significant on the 5% level. For the non-parametric Wilcoxon test, we do not find any significant return, though it should be noted that the p-value of 0.11 is very close to the threshold for significance at the 10% level. These results are in line with our predictions and support the argument that capital markets positively value the announcement of an M&A transaction in a strongly consolidated market, where suitable targets are becoming increasingly hard to find. However, a closer look at the other event windows suggests that the positive *CAARs* decrease the bigger the event window is defined eventually turning negative for the [-10;10] and [-20;20] event windows. While these results are not statistically significant, they suggest an interesting trend. Nonetheless, given the significant positive returns for the [-1;1] event window we conclude that the "big four" experience positive short-term value effects following the announcement of a transaction.

Table 3.13: CAARs to the "big four"

Total Sample of Transactions by the "big four" (N=36)								
Event-Window	CAAR	t-Test			z-Test			Wx Test	
		t-value	p-value		z-value	p-value		z-value	p-value
[-1; +1]	1.25%	2.04	0.04	**	1.99	0.05	**	-1.59	0.11
[-5; +5]	0.79%	0.96	0.34		1.12	0.27		-0.77	0.44
[-10; +10]	-0.81%	-0.72	0.48		-0.55	0.58		-0.55	0.58
[-20; +20]	-1.63%	-1.05	0.30		-0.85	0.40		-1.54	0.12

This table shows the cumulative average abnormal returns (CAAR) to the "big four" following the announcement of mergers and acquisitions in the brewing industry. It contains all transactions by the "big four" for which trading data was available between 250 before and 20 days after transaction announcement. Statistical significance at the 10%, 5% and 1% level is denoted by *, ** and *** respectively. The statistical significance has been tested using a standard t-test, the cross-sectional test as proposed by Boehmer, Musumeci, and Poulsen (1992) (z-Test) and the Wilcoxon Signed Rank Test as described by Barber and Lyon (1996).

Table 3.14 presents the rival returns to the "big four" i.e., the returns to the remaining 3 companies if one of the "big four" announces a transaction (e.g., the returns to Heineken, Carlsberg and SABMiller if Anheuser Busch Inbev announces a transaction). Using this approach we analyze 98 rival events for our sample. While we recognize a similar pattern of decreasing abnormal returns in case of larger event windows, we only find significantly negative returns for the [-10;10] event window. The returns are significant at the 10% level using a standard t-statistic and the non-parametric Wilcoxon signed rank test. The findings suggest that the negative competitive effects of missing out on a potential M&A opportunity or strengthened competition from a newly combined firm, outweigh potential positive signaling effects. These results stand in contrast to existing literature (see e.g., Eckbo, 1985; Song and Walkling, 2000; and Clougherty and Duso, 2009) and confirm the exceptional characteristics of the brewing sector.

Table 3.14: CAARs to the "big four" as Rivals

Total Sample of Rival Events (N=98)								
Event-Window	CAAR	t-Test		z-Test		Wx Test		
		t-value	p-value	z-value	p-value	z-value	p-value	
[-1; +1]	0.33%	1.02	0.31	1.09	0.28	-0.86	0.39	
[-5; +5]	0.25%	0.52	0.60	1.04	0.30	-0.26	0.79	
[-10; +10]	-1.32%	-1.92	0.06 *	-1.26	0.21	-1.69	0.09	*
[-20; +20]	-1.53%	-1.49	0.14	-0.64	0.52	-0.91	0.36	

This table shows the cumulative average abnormal returns (CAAR) to the "big four" following the announcement of mergers and acquisitions by a rival. It contains all events for which trading data was available between 250 before and 20 days after transaction announcement. Statistical significance at the 10%, 5% and 1% level is denoted by *, ** and *** respectively. The statistical significance has been tested using a standard t-test, the cross-sectional test as proposed by Boehmer, Musumeci, and Poulsen (1992) (z-Test) and the Wilcoxon Signed Rank Test as described by Barber and Lyon (1996).

Overall, our findings seem very consistent with previous industry research (see e.g., Kerkvliet et al., 1998; Earlam et al., 2010; and Schwankl, 2008) and confirm the importance of scale and synergies in the brewing sector. In a consolidated market, where it is becoming increasingly hard to find suitable M&A targets, capital markets seem to value the successful quest for a consolidation opportunity, while punishing rivals that miss out. Even though our study focuses solely on the brewing industry, the documented results may be indicative for other industries as well. In particular, industries with other consumer products and similar oligopolistic market structures (e.g., tobacco and breakfast cereals) may yield similar results.

3.5. Conclusion

The objective of this study was to analyze the short-term shareholder wealth effects of acquirers and rivals in the brewing industry. For this purpose, a sample of 69 horizontal M&A transactions involving brewing companies between 1998 and 2010 was identified and examined using a combination of two approaches: the traditional event study methodology and the Fama-French-3-Factor model. Our results provide new insights into the perceived short-term success of M&A transactions in the brewing industry and its corresponding evaluation through capital markets.

First, our results indicate that acquirers in the brewing industry experience significant positive short-term value effects following the announcement of an M&A transaction. This positive finding is an outstanding attribute of the sector and stands in contrast to cross-industry studies by Andrade, Mitchell, and Stafford (2001), Bruner (2002), and Loughran and Vijh (1997) and older industry specific research by Ebneth and Theuvsen (2007) all of which provide evidence of significant negative abnormal returns or at most insignificant positive returns. Therefore, it appears that capital markets specifically in recent years value the above-average synergy potential of the sector.

Second, our results provide evidence for a number of value drivers of abnormal acquirer performance. Consistent with the findings of Datta and Puia (1995) we find a positive relation between domestic transactions and short-term acquirer performance. However, at the same time, and in contrast to the findings of Williams and Liao (2008), we also find a positive impact of cross-border transactions involving targets in emerging markets, in particular Latin America, suggesting that capital markets value the international diversification strategies deployed by brewers. Moreover, in contrast to findings of Asquith, Bruner, and Mullins (1983), Jarrell and Poulsen (1989) and Moeller, Schlingemann, and Stulz (2003) we report a positive impact of acquirer size and transaction value, emphasizing the size advantages in the sector. On the other hand, our results provide evidence with regard to a negative relationship between acquirer returns and the target's public status. These results are consistent with Moeller, Schlingemann, and Stulz (2003) and Faccio, McConnell, and Stolin (2006), suggesting that brewers may have to pay significant premiums for targets that are publicly listed.

Third, our results indicate that the "big four" experience significant positive value effects upon announcing an M&A transaction suggesting that capital markets value the successful search for a suitable target in a strongly concentrated market. Close rivals, missing out on an M&A opportunity suffer from significant short-term value losses.

While our study addresses a number of important questions with regard to capital market effects of M&A in the brewing industry and the impact of determinant variables, the presented results also leave open questions and give rise to new research issues. As our study is limited to short-term capital market effects, the question arises whether acquiring brewers are able to sustain the positive announcement effects. Thus, future research could investigate the long-term implications of M&A in the sector. In addition to investigating capital market implications, future research could also analyze the impact of M&A on the operating performance of the involved companies. Given the expected continuation of the consolidation process, we believe that M&A in the brewing industry provide an interesting avenue for future research.

Furthermore, the results underline the importance of industry-specific M&A analyses, as sector related potentials to generate value differ among industries, resulting in possibly biased finding of cross-industry examinations.

4. Study 2: Long-term Success of Mergers and Acquisitions

4.1. Introduction

Towards the turn of the last century, many industries including the brewing industry have witnessed a significant increase in mergers and acquisitions (M&A) activity. Consolidation has been a major theme in the sector, as declining beer volumes in developed markets (in particular Western Europe), have forced many brewing companies to expand their operations into emerging markets in search for new growth opportunities. In addition to geographic diversification strategies, M&A have been further motivated by an extraordinary synergy potential of the brewing industry. In contrast to many other sectors the production, distribution and marketing of beer is characterized by a relatively high fixed cost base, resulting in high levels of operational leverage (Earlam et al., 2010) thus providing larger brewers with material size advantages. As a consequence, the global beer market today is dominated by four large multinational brewing groups: Anheuser-Busch Inbev, Heineken, SABMiller and Carlsberg, (the "big four"), who control in excess of 50% of global beer volumes. Despite rising market concentration, competition among the "big four" for market shares as well as potential acquisition opportunities has remained fierce (Iwasaki, Seldon, and Tremblay, 2008).

The shareholder wealth effects of M&A have been comprehensively discussed in empirical finance research. While Bruner (2002) concludes that the short-term wealth effects to acquirers average around zero, the majority of cross-industrial studies analyzing long-term performance provide consistent evidence of significant value losses to acquirers (e.g., Loughran and Vijh, 1997; Gregory, 1997; and Rau and Vermaelen, 1998). In addition to analyzing acquirer wealth effects, there is also growing interest in competitive effects to rival companies. Studies analyzing short-term rival effects provide evidence that rival companies gain at the M&A announcement due to positive information signaling effects (see Eckbo, 1983; Fee and Thomas, 2004; Shahrur, 2005; and Song and Walkling, 2000). In case of long-term event windows, Funke et al. (2008) conclude that capital markets do not immediately incorporate the effects of M&A into rival stock prices and report positive as well as negative intra industry effects.

In light of the specific industry characteristics of the brewing industry and the recent developments in the sector, Mehta and Schiereck (2011) analyze the short-term wealth effects of M&A in the brewing sector. The results of this study suggest that the global synergy and efficiency potential of M&A transactions is reflected by capital markets in the form of positive abnormal stock price reactions to acquiring companies. At the same time, the authors document significant negative returns to rivals missing out on potential M&A opportunities and identify a number of value drivers of abnormal performance. While the study provides new insights on short-term M&A success in the sector, the question, whether the documented effects are sustainable in the long run, remains open. Thus, the aim of this study is to fill this research gap and provide empirical evidence on the long-term effects of M&A in the sector. Our study determines the long-term performance of brewing companies based on a global dataset, using a combination of event and calendar time approaches to derive reliable indications of long-term performance, and analyses a comprehensive list of deal characteristics for their impact on the long-term wealth effects to acquiring companies. Moreover, our study specifically analyzes rival returns among the "big four" in the long run to show sector specific consolidation effects.

The remainder of this paper is structured as follows: Section 4.2 provides a brief overview of the relevant literature and outlines the derived hypotheses. Section 4.3 provides details on the applied methodology as well as the sample procedure. The following section 4.4 presents the empirical results and elaborates on the derived hypotheses. Finally, section 4.5 summarizes the findings and concludes.

4.2. Literature Review and Research Focus

Empirical finance research has extensively investigated the wealth effects of M&A. While the general consensus among researchers is that target companies gain significantly around the announcement of acquisitions (see e.g., Bradley, Desai, and Kim, 1988), the situation for acquiring companies is not as clear cut, with studies reporting significantly positive abnormal returns, significantly negative abnormal returns as well as abnormal returns that are not significantly different from zero (see Bruner, 2002). The evidence provided is usually based on event studies that use short-term event windows surrounding theannouncement date and implicitly assume that stock prices at once fully adjust and reflect the expected efficiency gains from acquisitions.

While the vast majority of research on wealth effects of M&A has focused on shareholder value effects surrounding the announcement date, i.e., the short-term wealth effects, a substantial body of literature also investigates the long-term wealth implications of M&A, challenging the assumption of market efficiency. The majority of these studies suggest a long-term underperformance of acquiring firms after engaging in M&A: Analyzing 937 mergers that took place between 1955 and 1987, Agrawal, Jaffe, and Mandelker (1992) report significantly negative abnormal acquirer returns over a five year post merger period. While Loughran and Vijh (1997) provide evidence that cash bidders experience significantly positive abnormal returns in the three to five year post-merger period, they also document significantly negative abnormal returns for the total sample. Similar results are reported by Gregory (1997) and Rau and Vermaelen (1998), who find significantly negative abnormal acquirer returns over a two-year and three-year post-merger period, respectively. While some researchers question the results of long-term studies, by pointing to mispricing issues and measurement problems (see e.g., Kothari and Warner, 2007), Agrawal and Jaffe (2000) conclude that overall, there seems to be enough evidence to suggest an anomaly following mergers.

Empirical research also investigates the impact of M&A on rival firms. With regard to short-term wealth effects, findings show positive as well as negative effects on rivals: On the one hand, rival companies may benefit from the transaction announcement as a result of a positive signaling effect regarding industry attractiveness and potential future takeover activity (Eckbo, 1983; Song and Walkling, 2000). Moreover, a merger in the industry decreases the number of competitors and thus increases the likelihood of collusion, which may lead to greater monopoly rents to rival firms (Eckbo, 1983; Shahrur, 2005). On the other hand, rival firms may be impacted by negative competitive effects due to more intense competition in the industry from a new, more efficient combined firm (Eckbo, 1983). Overall, the positive effects outweigh the negative competitive effects: For example Eckbo (1985) reports positive announcement effects to rivals for horizontal transactions. Similar results are found by Song and Walkling (2000) in a study including horizontal and non-horizontal transactions. These results are confirmed in more recent studies by Clougherty and Duso (2009), Fee and Thomas (2004) and Shahrur (2005). While the above-mentioned studies focus on short-term rival effects, Funke et. al. (2008) analyze the long-term intra industry effects utilizing a sample of 2,511 transactions from 1985 to 2005. The study provides evidence of rival gains due to positive information signaling as well as rival losses due to negative competitive effects and hence concludes that capital markets do not immediately incorporate the effects of M&A into rival stock prices.

As mentioned above the structure of the global brewing industry has significantly changed over the course of the last decade. The brewing sector has been and still remains particularly prone to consolidation and M&A, as in contrast to many other sectors, increased size provides brewers with a material competitive advantage. Given the particularities of the industry, it seems only logical to conduct industry-specific M&A analyses in order to account for sector-specific potentials. Nonetheless, empirical studies on the brewing industry remain scarce and are mostly limited to the US brewing industry. The specific topics addressed in the studies focus on technological change in the sector (Kerkvliet et al., 1998), its tendencies towards concentration (Lynk, 1985; Adams, 2006), the determinants and motives for horizontal M&A (Tremblay and Tremblay, 1988), as well as competition in the industry (Horowitz and Horowitz, 1968). More recently, Ebneth and Theuvsen (2007) analyze the short-term value effects of M&A to acquirers using event study methodology. Based on a sample of 29 cross-border transactions involving European acquirers from 2000-2005, they observe positive acquirer returns, which lack significance though. In a companion study comprising 69 M&A transactions, Mehta and Schiereck (2011) challenge these results and report significant positive abnormal returns to acquirers. Moreover, they find a significant negative effect to close rivals missing out on a potential M&A opportunity.

With this paper we aim to contribute to existing literature in two ways. First, we provide a comprehensive analysis of long-term wealth effects of M&A to acquirers in the brewing industry using a global dataset that in addition to cross-border transactions also includes domestic transactions. This enables us to specifically build and investigate relevant subsamples and determine key drivers of abnormal performance. Second, we specifically analyze transactions by the "big four" as well as the competitive effects on rivals missing out on a transaction.

Our main research is concentrated on acquirers' long term performance and its determinants as well as the corresponding rival effects of the "big four".

i) Acquirer Long-term Performance:

Given the particularities of the brewing industry and its development over the last few years, M&A in the sector may be seen as a feasible measure to realize synergy and efficiency gains. In a companion paper Mehta and Schiereck (2011) analyze 69 M&A transactions in the sector and document significant positive announcement effects to acquiring brewers. They explain these results with the extraordinary synergy potential perceived by capital markets. Despite the fact that previous cross-industrial studies examining the long-term effects of M&A provide consistent evidence of long-term value losses to acquirers, we assume acquirers to sustain the positive announcement effects. Consequently, we expect to find significantly positive abnormal long-term returns to acquirers.

In order to determine potential drivers of long-term performance, a number of deal, acquirer, and target characteristics are examined.

ii) Determinants of Acquirer Performance:
While various studies investigate the impact of geography on acquirer performance, most of them are limited to short-term analyses. Nevertheless, several studies also examine the effect of internationalization on long-term performance. The results of these studies provide consistent evidence that suggest a negative relationship between cross-border transactions on long-term acquirer performance: After investigating 4,000 domestic and cross-border transactions by UK-based bidding firms, Conn et al. (2005) come to the conclusion that cross-border transactions result in significantly lower long-term returns than domestic transactions. Similar results are documented by Aw and Chatterjee (2004) and Black, Carnes, and Jandik (2001). It is argued that the integration of overseas targets is likely to result in a more challenging post-merger integration due to cultural differences of the merging firms. Moreover, there are difficulties in accurately valuing overseas targets due to imperfect capital information in countries with less developed capital markets.

In case of the brewing industry, cross-border transactions have been of particular relevance as many brewers have looked to diversify into international markets, in particular emerging markets, in an attempt to offset the continuous decline in beer volumes in many mature markets. In line with previous research we expect to find significantly negative long-term returns for cross-border transactions. On the other hand, we assume acquirer returns to be less negatively impacted if targets are based in emerging market economies, as we believe growth opportunities may offset potential integration issues.

The impact of transaction value on long-term acquirer performance has been less focused on in previous research. Existing industry-specific studies suggest a positive relationship between transaction value and long-term performance: Ferris and Park (2002) report a positive impact of transaction value on acquirer returns in the telecommunications industry. Similarly, Laabs and Schiereck (2010) confirm these results for the automotive supplier industry. They explain these findings with potentially greater economies of scale for larger targets. On the other hand, the integration of larger targets may be more difficult than for small targets (Hawawini and Swary, 1991). In case of the brewing industry, increased company size provides brewers with a material competitive advantage. Hence, it can be argued that the acquisition of large targets will significantly contribute to the success of a transaction due to greater potential for revenue and cost synergies. Thus, in line with evidence from other industries, we expect to find a positive impact of transaction value on long-term acquirer performance.

The acquisition of privately held companies accounts for the majority of M&A transactions in the brewing industry. The general consensus among researchers investigating short-term acquirer returns is that bids for privately held companies generate higher bidder returns than bids for publicly held companies (see e.g., Moeller, Schlingemann, and Stulz, 2003; and Faccio, McConnell, and Stolin, 2006). Martynova and Renneboog (2006) argue that, in the case of privately held targets, bidders are likely to benefit from price discounts as compensation for buying a comparably illiquid stake. At the same time, they see advantages due to the fact that private companies usually have fewer shareholders, which facilitates negotiations. Consequently, we expect to find significant differences in the long-run performance of acquirers of private and public targets.

Several studies investigate the impact of the method of payment on acquirer performance. Myers and Majluf (1984) argue that bidders prefer to pay using stock when they believe that the market overvalues their shares and on the other hand prefer using cash, when they regard their stock as undervalued. Similarly, Martynova and Renneboog (2006) suggest that the means of payment used is an important signal of the quality of the target firm and its potential synergy value. They argue that a cash offer by the bidding company signals a willingness to pay off target shareholders in order to avoid sharing future cash flows and bear the sole risk of the combined firms. On the other hand, an all-equity offer signals the willingness to keep the target shareholders involved in the merged company and share its risk. The theoretical framework behind the mentioned signaling effects is supported by the results of both short-term studies (see e.g., Brown and Ryngaert, 1991; and Wansley, William, and Yang, 1983) and long-term studies (see Loughran and Vijh, 1997), all of which provide evidence of significantly positive acquirer returns for cash transactions and significantly negative acquirer returns in case of all-share transactions. We expect to find similar results for the brewing industry.

Over the last few years the brewing industry has experienced a sharp increase in M&A activity and seen strong industry consolidation. As a consequence, the competitive landscape has materially changed as the "big four" today control more than 50% of global beer volumes. While the sector is expected to further consolidate, it is becoming increasingly difficult to find suitable targets (Gibbs, Webb, and Dhillon, 2010). We assume the changes in market structures and concentration to have an impact on acquirer performance and expect to find significant differences in returns over time.

iii) Wealth and Rival Effects Among the "big four":

As already mentioned above, the global beer market is dominated by Anheuser-Busch Inbev, SABMiller, Heineken and Carlsberg (the "big four"). Mehta and Schiereck (2011) investigate short-term announcement effects and specifically focus on rival effects among the "big four". They find significantly positive abnormal returns if one of the "big four" announces an M&A acquisition, while the remaining three rivals, missing out on an M&A opportunity experience significantly negative abnormal returns. Mehta and Schiereck (2011) explain these results by strong competition among the "big four" and the increased difficulty to find suitable targets in a fairly concentrated market. Specifically, capital markets seem to value the successful search for an M&A target, while rivals missing out are punished. We expect to find a similar effect with regard to long-term abnormal performance.

4.3. Data Selection and Research Methodology

4.3.1. Data Selection

The sample of mergers and acquisitions for the long-term event study is drawn from the Securities Data Corporation (SDC)/ Thomson One Banker Deals database and the Merger Market M&A database. It includes all world-wide M&A events announced between January 1st, 1998 and December 31st, 2010. The total number of M&A deals is reduced to yield only those transactions meeting the following criteria:

i.   At the time of the transaction, acquirer and target companies both had active operations in the brewing industry.

ii. The acquiring company has been publicly listed for at least 250 days prior to the announcement of the transaction.

iii. The total transaction value accumulates to at least USD 50 million.

iv. The completion of the transaction leads to a change of control in the target; prior to the announcement of the transaction the bidder holds less than 50% in the target company, following the transaction the bidder obtains a controlling stake in the target company.

v.   The transaction has been successfully completed.

In addition, the transactions were validated by a press research using the Factiva database as well as company websites in order to ensure that all transactions are horizontal and the announcement dates provided by the databases are correct. Moreover, acquirers with multiple transactions on the same day as well as acquirers with limited available trading data were removed from the dataset. The described selection criteria result in a final sample of 66 transactions. The frequency distribution of the transactions over time is provided in Table 4.1. While the number of transactions is spread fairly even from 1998 onwards, the average transaction size varies strongly from USD 382 million to USD 11,973 million due to a number of high-profile transactions such as InBev's acquisition of Anheuser-Busch (USD 52 billion), Heineken and Carlsberg's takeover of S&N (USD 15 billion) and Heineken's recent acquisition of Femsa Cerveza (USD 5.7 billion). In terms of geography more than 80% of the transactions involve acquirers that are based in Europe.

The relevant daily stock prices, market capitalizations and local market indices for acquirers were downloaded from the Thomson Datastream database. Acquirer returns are calculated using the Datastream Total Return Index, which adjusts the closing share prices for dividend payments as well share issuances or repurchases.

Table 4.1: Sample Overview: Descriptive Statistics

Year	Deals	(%)	Avg. Trans. Val. (USD mil.)	Trans. Val. (USD mil.)	Acquirer Region - Number of Deals			
					Europe	Americas	Asia	RoW
2010	1	1.5	5,700	5,700	1			
2009	4	6.1	737	2,946	2		2	
2008	6	9.1	11,973	71,835	6			
2007	5	7.6	405	2,025	1	2	1	1
2006	5	7.6	382	1,909	5			
2005	8	12.1	785	6,281	8			
2004	8	12.1	923	7,385	6	1	1	
2003	5	7.6	671	3,357	5			
2002	6	9.1	1526	9,157	5	1		
2001	4	6.1	599	2,396	3	1		
2000	6	9.1	514	3,086	5		1	
1999	4	6.1	495	1,981	3	1		
1998	4	6.1	268	1,071	4			
Sum	**66**	**100.0**	**1,805.0**	**119,129.0**	**54**	**6**	**5**	**1**

Sample includes all M&A transactions between 1998 and 2010 as specified above. Table 1 shows the frequency distribution of all M&A transactions between 1998 and 2010 with total and average transaction values in USD mil. Additionally, details on acquirer region are provided.

4.3.2. Research Methodology

Existing research assessing long-term performance utilizes two distinct methodologies to measure abnormal performance: The buy-and-hold-abnormal-return (BHAR) approach and the calendar time portfolio approach. Even though there is extensive literature comparing both methodologies, there is no clear superiority of one over the other (see e.g., Kothari and Warner, 2007; Barber and Lyon, 1997; and Mitchell and Stafford, 2000). Thus, in order to provide for an accurate and robust assessment of long-term performance we analyze long-term performance using a combination of both approaches.

The *BHAR* approach goes back to the 1990s, when it was introduced by Ritter (1991) in a study analyzing the long-term performance of Initial Public Offerings (IPOs). Since then the *BHAR* methodology has been applied in a number of long-term event studies and further advanced (Loughran and Vijh, 1997). The *BHAR* methodology is also referred to as the characteristic-based matching approach, as it involves matching event firms with non-event firms based on stock characteristics such as market capitalization and market-to-book ratios. Abnormal returns are then calculated as the difference in stock performance of event firms and the matched non-event benchmark firms over a specified period. Lyon, Barber, and Tsai (1999) argue that *BHARs* more accurately represent an investor's actual investment behavior than calendar time portfolios, which assume a monthly rebalancing of investor portfolios. Moreover, Loughran and Ritter (2000) provide evidence that the *BHAR* methodology has substantially higher statistical power to detect abnormal returns when compared to its counterpart. On the other hand, critics of the *BHAR* approach argue that it is exposed to significant cross-correlation, especially due to its use of matching samples that are not randomly selected (see Mitchell and Stafford, 2000).

The *BHAR* approach requires a set of non-event firms from the same industry that are used as benchmark/control firms. Since there is no broad global index available for the brewing industry, we determine a portfolio of matching firms based on the constituents' lists underlying the country-specific industry indices for companies in the brewing industry. After downloading and screening all available lists as supplied by the Thomson Datastream database and verifying the results with the Merger Market database, we count a total number of 73 publicly listed matching firms. For each firm monthly stock-returns as well as monthly market values and yearly market to book ratios are downloaded from the Thomson Datastream database.

We determine *BHARs* using a character-based matching approach on the basis of market values and market-to-book ratios as proposed by Lyon, Barber, and Tsai (1999):

i. For each acquiring firm in the sample market values and market-to-book ratios are determined. Market values are determined as the last quote in the last June preceding transaction announcement; market-to-book ratios are derived for the last completed fiscal year before transaction announcement.

ii. Likewise, for each year, market values at the end of June and market-to-book ratios are downloaded for the full list of matching firms.

iii. In a second step, the list of potential matches is reduced to those companies with a market value range of 70% to 130% of the acquirer's market value.

iv. Finally, from the list of companies with a market value between 70% and 130% of the acquirer's market value, the one with the smallest absolute difference in market-to-book value is selected as the control firm for the analysis.

Based on this procedure, we determine control firms, i.e., the benchmark reference, for each acquiring brewer and rival as part of our analysis of competitive effects. The abnormal returns are then derived as the difference between the Buy-and-Hold Return (BHR) of an investor in the event company and the BHR of the control firm (see equation 4.1). The average BHARs are then calculated as an equal-weighted average. Statistical significance is tested using a standard t-statistic. In order to compare mean differences in subsamples, we utilize the two groups difference of means test proposed by Cowan and Sergeant (2001).

$$BHAR_{i,[t,T]} = \prod_{t=1}^{T}\left(1 + R_{i,t}\right) - \prod_{t=1}^{T}\left(1 + R_{j,t}\right) \qquad (4.1)$$

The calendar time methodology was introduced by Jaffe (1974) and Mandelker (1974) and is particularly supported by Fama (1998) and Mitchell and Stafford (2000). In general the methodology involves creating calendar time portfolios of event firms. For each month of the sample, a portfolio is constructed including all firms that have experienced an event in the analyzed time window. Due to the fact that the number of firms in a portfolio changes over time an equal weighted portfolio return is calculated for every month in the sample period. While the resulting time series of portfolio returns is free of any cross-correlation issues and can thus be used for standard statistical tests, it should be noted that portfolio returns may be affected by other potential pitfalls such as heteroscedasticity (Mitchell and Stafford, 2000).

In order to determine abnormal performance, the calendar-time portfolio approach is usually combined with a multi-factor asset pricing model. In our study we utilize the Fama-French-3-Factor-model (FF3F). Specifically, we regress the time-series of calendar time portfolio excess returns with the three Fama and French (1992) market, size (SMB), and book-to-market (HML) factors as presented in equation 4.2.

$$R_{i,t} - R_{f,t} = \alpha_i + b_i\left(R_{M,t} - R_{f,t}\right) + s_i SMB_t + h_i HML_t + e_{i,t} \qquad (4.2)$$

The time series of SMB and HML are determined as proposed by Faff (2003) utilizing style indices by data provider Russel. Inferences about abnormal performance can then be drawn on the basis of the estimated intercept α, which represents the average monthly abnormal return to the observed acquirers over the observed time period following an acquisition. Statistical significance is determined using a standard t-statistic.

4.4. Empirical Results

In the following the empirical results of our analyses are presented. We start off reporting the results for the total acquirer sample, which is followed by univariate and multivariate analyses, in order to determine potential drivers of abnormal performance. Moreover, we separately report wealth effects among the "big four".

4.4.1. Overall Acquirer Performance

Table 4.2 provides an overview of the long-term buy-and-hold-abnormal-returns (BHARs) for the total sample of acquiring companies from 6 month to 24 months following transaction announcement. Overall, the calculated BHARs are statistically insignificant throughout all examined event periods and range from 1.26% for a 12 month holding period to -4.63% for a 24 month holding period. The results stand in contrast to our expectations and suggest that overall brewers cannot sustain the previously reported positive announcement effects documented by Mehta and Schiereck (2011).

Table 4.2: BHARs to Acquirers

Holding Period	6 Months	12 Months	18 Months	24 Months
BHAR	-2.10%	1.26%	-1.49%	-4.63%
t-value	-0.79	0.33	-0.29	-0.75
p-value	0.43	0.74	0.78	0.45
N	66	66	65	62

This table shows the average Buy-and-Hold Abnormal Returns (BHAR) to acquiring companies in mergers and acquisitions in the brewing industry. Abnormal returns are derived using a control-firm matching approach as proposed by Lyon, Barber, and Tsai (1999). All acquirers between 1998 - 2010 are included for which the relevant matching information is available (market values and market-to-book ratios). Statistical significance at the 10%, 5%, and 1% level is denoted by *, **, *** respectively, and is tested using a standard t-test.

In order to investigate the robustness of the above results we analyze calendar portfolio returns as an alternative methodology to the use of BHARs. Table 4.3 documents the results of applying a FF3F model, as specified in section 3, on the previously determined calendar portfolio returns. Overall, all of the analyzed models are statistically significant and have relatively high explanatory power. We report positive abnormal monthly returns throughout all examined holding periods ranging from 0.11% to 0.62%. However, none of the alphas is statistically significant and hence there is no conclusive evidence of a significant abnormal long-term performance. In addition to the regression intercepts, Table 4.3 also reports the factor loadings of the portfolios. The results generally suggest that the analyzed portfolios represent stocks with relatively low risk with respect to the FF3F model: All of the analyzed portfolios have market betas that are below one and significantly positive loadings on the value factor with negligible exposure to the size factor.

Overall, the results of both applied methodologies suggest that acquiring companies in the brewing industry do not experience significant abnormal long-term wealth effects. In order to gain further insights into potential determinants and drivers of abnormal performance, in the following, we investigate a number of deal, acquirer and target characteristics using subsamples.

Table 4.3: Acquirer Calendar Time Portfolio Returns

Holding Period	6 Months		12 Months		18 Months		24 Months	
Average raw return	0.90%		1.48%	***	1.15%	**	1.28%	***
t-value	1.32		2.75		2.29		2.65	
p-value	0.19		0.01		0.02		0.01	
Fama French three factor alphas								
Alpha	0.33%		0.62%		0.11%		0.14%	
t-value	0.52		1.43		0.28		0.34	
p-value	0.60		0.07		0.78		0.73	
beta market	0.97	***	0.93	***	0.84	***	0.80	***
t-value	8.77		12.20		11.57		11.42	
p-value	0.00		0.00		0.00		0.00	
beta SMB	-0.13		0.19		0.17		0.28	*
t-value	-0.50		1.08		0.98		1.69	
p-value	0.62		0.28		0.33		0.09	
beta HML	0.69	***	0.60	***	0.59	***	0.53	***
t-value	3.36		4.16		4.29		4.00	
p-value	0.00		0.00		0.00		0.00	
R-squared	0.35		0.51		0.49		0.49	
Adjusted R-squared	0.34		0.50		0.48		0.48	
f-statistic	25.81	***	51.97	***	46.67	***	46.84	***
p-value	0.00		0.00		0.00		0.00	
DW	2.17		2.39		2.20		2.16	

This table shows the results of a long-term event study, based on the Fama-French-3-Factor-model, for the total sample of acquiring companies in mergers and acquisitions in the brewing industry. The intercept "alpha" stands for the average monthly abnormal return to the acquiring companies over the specified holding period. All acquirers between 1998 -2010 are included. Statistical significance at the 10%, 5%, and 1% level is denoted by *, **, *** respectively, and is tested using a standard t-test. In addition, statistical quality of the model is indicated by the adjusted determination coefficient "Adjusted R-squared" and the corresponding f-statistic.

4.4.2. Analysis of Determinants of Acquirer Performance

Geographical Scope of Transaction: Tables 4.4 and 4.5 present the findings regarding the impact of geographic diversification on long-term acquirer performance. Overall, acquirers show a preference for cross-border transactions and in particular emerging market transactions. In our total sample of 66 transactions, only 29% (19 transactions), are domestic/national transactions, while 71% (47 transactions) are cross-border transactions. Approximately 64% of these cross-border transactions qualify as emerging market transactions and involve targets that are based in Latin-America, Asia (ex Japan) or Eastern Europe.

On average domestic acquirers experience positive value effects in every analyzed event window with *BHARs* ranging between 3.63% and 23.61%. Despite the small sample size of only 19 transactions *BHARs* are significant at the 5% and 1% level for a 12 month and 24 month holding period, respectively. On the other hand, we consistently report negative *BHARs* for cross-border transactions across all analyzed holding periods ranging from -4.41% over a 6 month holding period to a significantly negative -16.18% over a 24 month holding period. A comparison of means shows that *BHARs* for cross-border transactions are significantly lower for almost every analyzed event window. These results are in line with our expectations and cross-industrial studies by Conn et al. (2005), Aw and Chatterjee (2004) and Black, Carnes, and Jandik (2001). In addition, the results will be challenged in the multivariate analysis, not least due to the relatively small amount of domestic transactions.

Table 4.5 compares domestic transactions with emerging market transactions. Overall, *BHARs* for emerging market transactions are insignificant throughout all analyzed event windows, ranging from -11.47% to 0.17%. A comparison of means shows that *BHARs* for emerging market transactions are significantly lower than for domestic transactions in case of a 24 month holding period. On the other hand, we do not find significant differences between cross-border transactions in mature and emerging market transactions (see Table A4.1 in Appendix). Again, due to the limited amount of transactions the results will be challenged in the multivariate analysis.

Table 4.4: BHARs by Geographical Scope

	Holding Period	6 Months	12 Months		18 Months	24 Months	
Domestic	BHAR	3.62%	18.51%	**	13.67%	23.61%	***
	t-value	0.66	2.28		1.33	2.72	
	p-value	0.51	0.03		0.20	0.01	
	N	19	19		19	18	
Cross-border	BHAR	-4.41%	-5.72%		-7.74%	-16.18%	**
	t-value	-1.48	-1.51		-1.33	-2.22	
	p-value	0.15	0.14		0.19	0.03	
	N	47	47		46	44	
Difference	Delta	-8.03%	-24.22% ***		-21.41% *	-37.79%	***
	t-value	-1.29	-2.70		-1.82	-3.52	
	p-value	0.20	0.01		0.07	0.00	

This table shows the average Buy-and-Hold Abormal Returns to acquiring companies in mergers and acquisitions in the brewing industry differentiated by geographical scope. Abnormal returns are derived using a control-firm matching approach as proposed by Lyon, Barber, and Tsai (1999). All acquirers between 1998 - 2010 are included for which the relevant matching information is available (market values and market-to-book ratios). Statistical significance at the 10%, 5%, and 1% level is denoted by *, **, *** respectively, and is tested using a standard t-test. Statistical significance of mean differences is tested using the two groups difference of means test as proposed by Cowan and Sergeant (2001).

Table 4.5: BHARs by Geographical Scope

	Holding Period	6 Months	12 Months		18 Months	24 Months	
Domestic	BHAR	3.62%	18.51%	**	13.67%	23.61%	***
	t-value	0.66	2.28		1.33	2.72	
	p-value	0.51	0.03		0.20	0.01	
	N	19	19		19	18	
Emerging Market	BHAR	0.17%	3.12%		0.33%	-11.47%	
	t-value	0.05	0.53		0.04	-1.17	
	p-value	0.96	0.60		0.97	0.25	
	N	30	30		29	28	
Difference	Delta	3.45%	15.39%		13.34%	35.08%	***
	t-value	0.52	1.53		1.05	2.67	
	p-value	0.60	0.13		0.30	0.01	

This table shows the average Buy-and-Hold Abormal Returns to acquiring companies in mergers and acquisitions in the brewing industry differentiated by geographical scope. Abnormal returns are derived using a control-firm matching approach as proposed by Lyon, Barber, and Tsai (1999). All acquirers between 1998 - 2010 are included for which the relevant matching information is available (market values and market-to-book ratios). Statistical significance at the 10%, 5%, and 1% level is denoted by *, **, *** respectively, and is tested using a standard t-test. Statistical significance of mean differences is tested using the two groups difference of means test as proposed by Cowan and Sergeant (2001).

Impact of Transaction Size: In order to investigate the impact of transaction size on long-term performance, the total transaction sample is ranked by the underlying transaction value in USD. We then build subsamples with the 30 largest and 30 smallest transactions and assess differences in performance. The results are documented in Table 4.6. On average, acquirers in large transactions yield negative BHARs ranging from -2.02% to -11.26%. In case of a 6 month holding period acquirers yield a negative -6.64% which is significant at the 10% level. On the other hand, BHARs for small transactions are statistically insignificant, ranging from -0.88% to 1.79%. On average, BHARs for small transactions are consistently higher for small transactions across all analyzed event windows. However, the differences in performance are not statistically significant.

Table 4.6: BHARs by Transaction Size: Top 30 vs. Bottom 30

Holding Period		6 Months		12 Months	18 Months	24 Months
Top 30	BHAR	-6.64%	*	-2.02%	-6.26%	-11.26%
	t-value	-1.98		-0.46	-1.28	-1.44
	p-value	0.06		0.65	0.21	0.16
Bottom 30	BHAR	-0.88%		0.53%	0.43%	1.79%
	t-value	-0.22		0.09	0.04	0.16
	p-value	0.83		0.93	0.96	0.87
Difference	Delta	-5.94%		-3.21%	-8.84%	-15.30%
	t-value	-1.16		-0.43	-0.83	-1.15
	p-value	0.25		0.67	0.41	0.26

This table shows the average Buy-and-Hold Abormal Returns to acquiring companies in mergers and acquisitions in the brewing industry differentiated by transaction size. Abnormal returns are derived using a control-firm matching approach as proposed by Lyon, Barber, and Tsai (1999). All acquirers between 1998 - 2010 are included for which the relevant matching information is available (market values and market-to-book ratios). Statistical significance at the 10%, 5%, and 1% level is denoted by *, **, *** respectively, and is tested using a standard t-test. Statistical significance of mean differences is tested using the two groups difference of means test as proposed by Cowan and Sergeant (2001).

Impact of Public Status of Target: Table A4.2 compares the BHARs of transactions involving publicly listed targets and privately held targets. The acquisitions of publicly listed targets consistently yield negative BHARs ranging from -0.90% to -10.83%, all of which are statistically insignificant. The BHARs for acquisitions of privately held targets are on average slightly higher in comparison ranging from -1.73% to 2.84%, but are also insignificant. A comparison of means shows no significant differences in BHARs to confirm a potential relationship between the public status of the target and acquirer abnormal performance. Nonetheless, we will reinvestigate the issue in the multivariate analysis.

4.4.3. Wealth and Rival Effects among the "big four"

While the above analyses covered transactions of all publicly listed acquirers, in the following we give particular attention to wealth effects of the "big four" and rival effects among them i.e. the wealth effect to the remaining three companies if one of the "big four" announces a transaction (e.g., the BHARs to Heineken, Carlsberg and SABMiller if Anheuser Busch Inbev announces a transaction). Table 4.7 presents the BHARs of acquisitions by the "big four". On average, the reported BHARs are negative across all analyzed event windows ranging from -3.27% to -11.65%. However, none of the returns is statistically significant. Furthermore, we also do not find any significant differences in BHARs of acquisitions by the "big four" and acquisitions by other acquirers (see A4.3 in Appendix). Nonetheless, overall, the results imply that the "big four" cannot sustain the positive short-term announcement returns as determined by Mehta and Schiereck (2011).

Table 4.7: BHARs: Acquisitions by "big four"

	Holding Period	6 Months	12 Months	18 Months	24 Months
"big four"	BHAR	-3.90%	-3.27%	-4.06%	-11.65%
	t-value	-1.05	-0.77	-0.70	-1.28
	p-value	0.30	0.45	0.49	0.21
	N	35	35	34	34

This table shows the average Buy-and-Hold Abnormal Returns for transactions by the "big four" (Anheuser-Busch Inbev, SABMiller, Heineken and Carlsberg). Abnormal returns are derived using a control-firm matching approach as proposed by Lyon, Barber, and Tsai (1999). All acquirers between 1998 - 2010 are included for which the relevant matching information is available (market values and market-to-book ratios). Statistical significance at the 10%, 5%, and 1% level is denoted by *, **, *** respectively, and is tested using a standard t-test.

Table 4.8 presents the rival BHARs of the "big four" when missing out on a transaction. Overall, we report ambiguous results with regard to rival returns. While we document insignificant negative rival BHARs over a 6 month and 12 month holding period, we find significantly positive rival BHARs of 11.22% and 23.96% over an 18 month holding period and 24 month holding period, respectively. In order to assess the robustness of these results, we additionally analyze calendar portfolio returns, which are displayed in Table A4.4 in the appendix. The results of the calendar time approach do not confirm the positive abnormal returns as suggested by the BHAR analysis. Nonetheless, there is sufficient evidence to conclude that the negative short-term rival effects among the "big four", as documented by Mehta and Schiereck (2011), are not sustainable in the long-run.

Table 4.8: BHARs: "big four" Rival Returns

	Holding Period	6 Months	12 Months	18 Months	24 Months
Rivals	BHAR	-2.56%	-2.19%	11.22% **	23.96% ***
	t-value	-1.15	-0.54	2.41	4.58
	p-value	0.26	0.59	0.02	0.00
	N	78	78	78	78

This table shows the average Buy-and-Hold Abnormal Returns to the "big four" (Anheuser-Busch Inbev, SAB-Miller, Heineken and Carlsberg), when missing out on an acquisitions opportunity to one of the other three rivals. Abnormal returns are derived using a control-firm matching approach as proposed by Lyon, Barber, and Tsai. (1999). All acquirers between 1998 - 2010 are included for which the relevant matching information is available (market values and market-to-book ratios). Statistical significance at the 10%, 5%, and 1% level is denoted by *, **, *** respectively, and is tested using a standard t-test.

4.4.4. Multivariate Regression Analysis

In order to provide a complete picture of the influential factors and to gain further insights into potential dependencies, a cross-sectional regression is performed on the buy-and-hold-abnormal-returns to acquirers as presented in equation 4.3. In total, seven variables are included in the regression model to represent the parameters, which have been individually analyzed in the univariate subsample analysis. In addition, we include two variables to assess the impact of the type of consideration and time. In the following, we specify the respective parameter values in detail.

$$BHAR_n = \alpha_0 + \delta_1 * CB + \delta_2 * EM + \delta_3 * TV + \delta_4 * PT + \delta_5 * B4$$
$$+\delta_6 * SC + \delta_7 * D_A$$

$$(4.3)$$

Geographical scope: The results presented in the univariate analysis provide a first indication that cross-border transactions might have a negative impact on long-term acquirer performance, when compared to domestic transactions. At the same time, we did not find significant negative returns for cross-border transactions where targets were based in emerging market economies (Latin America, Eastern Europe and Asia (ex Japan)). We reinvestigate both effects in the regression models using the dummy variables CB and EM.

Transaction size: While not statistically significant, the BHARs for large transactions were on average lower than for small transactions in every analyzed event window. In order to verify the univariate results, we include transaction values in the regression model (TV).

Public Status of target: The univariate results did not provide sufficient evidence to explain a potential effect of the target's public status on long-term performance. We further investigate the issue by including a dummy variable in the regression models for publicly listed targets (PT).

"big four" transaction: The univariate analysis suggested no significant long-term returns to the "big four" and no significant differences to acquisitions by other acquirers. In order to verify these results we include a dummy variable for transactions by the "big four" ($B4$).

Type of consideration: The impact of the type of consideration used in transactions was not analyzed in the univariate analysis. In order to analyze a potential relationship between the type of consideration used i.e., cash/share based, we include a dummy variable for transactions that are partly or completely financed with shares (SC).

Time period: As pointed out in section 2, the structure of the global beer market has materially changed in recent years. In order to test for a supposed relation between time and acquirer performance, we include a dummy variable for time period 1998 and 2003(D_A).

Table 4.9: Multivariate Regression on Acquirer 12 month BHAR

Dependent variable: Acquirer 12 month *BHAR*		M1	M2	M3	M4
Constant	α	0.10 *1.08*	0.14 * *1.77*	0.10 *1.38*	0.16 ** *2.43*
Independent variables					
Cross-border M&A	δ_1	-0.21 * *-1.94*	-0.18 ** *-2.18*	-0.18 ** *-2.21*	-0.21 *** *-2.66*
Target emerging market	δ_2	0.02 *0.23*			
Transaction value	δ_3	0.00 *1.10*			
Public target	δ_4	-0.11 *-1.42*	-1.02 *-1.41*		
"big four" transaction	δ_5	-0.03 *-0.24*			
Share-comp	δ_6	*0.18* ** *2.00*	*0.16* * *1.88*	*0.14* * *1.70*	
1998 – 2003	δ_7	0.04 *0.53*			
R² (adj.)		0.09	0.13	0.11	0.09
F-stat. (p-value)		0.09 *	0.01 **	0.00 ***	0.0 ***
DW-statistic		1.69	1.58	1.56	1 1.62

This table shows the results of a cross-sectional regression analysis of 12 month acquirer Buy-and-Hold Abnormal Returns (BHAR) on a host of explanatory variables. ***, **, and * indicate statistical significance at the 1%, 5%, and 10% levels, respectively. t-statistics are presented in italics.

Table 4.9 presents the results of four regression models (M1-M4) on the BHARs for a 12 month event window. The regression models differ with regard to the independent variables included. In general, all of the regression models are statistically significant (based on F-statistics) and explanatory power is remarkably high with adjusted R-squared ranging between 9% and 13%. Autocorrelation issues can be ruled out as Durbin-Watson-Statistics are fairly close to two. The overall average abnormal long-term return as represented by the constant is positive throughout all regression models. Moreover for M2 and M4 the reported returns of 14.00% and 16.00% are significant at the 10% and 5% level, respectively. While these results are not consistent with the univariate BHAR analysis and the results of the calendar time portfolio approach, they are in line with our expectations and confirm the specific characteristics of the sector. In contrast to existing cross-industrial studies, which provide consistent evidence of long-term value losses to acquirers, our analysis suggests that returns to acquirers in the brewing industry are at most insignificant if not significantly positive. These results represent an outstanding attribute of the sector and suggest that capital markets value the extraordinary synergy potential.

In addition, the regression models complement the univariate subsample analysis enabling the detection of potential value drivers of long-term performance. The multivariate analysis confirms the negative impact of cross-border transactions on acquirer performance as coefficients are significant across all four models. This finding confirms our expectations and is in line with existing cross-industrial studies (Conn et al., 2005; Aw and Chatterjee, 2005; and Black, Carnes, and Jandik, 2001). On the other hand, regression coefficients for emerging market transactions are insignificant, suggesting that potential integration issues may be offset by growth opportunities in emerging markets.

The univariate analysis suggested no significant difference in long-term performance between large and small transactions. The multivariate regression analysis does not provide any additional insight as we report an insignificant positive regression coefficient.

With regard to the impact of the target's public status on acquirer performance, the univariate analysis suggested no significant difference between public and private targets. These results are confirmed by the regression analysis, as we find insignificant negative coefficients for public acquisitions.

The univariate analysis suggested no significant difference in long-term performance of acquisitions by the "big four" and other acquirers. These results are confirmed by the regression models as we do not find a significant coefficient.

While the majority of transactions in the total sample are cash transactions, a number of transactions include share based compensation or are entirely financed with shares. Contrary to the findings of Loughran and Vijh (1997), the results of the regression analyses provide some evidence of a positive impact of share-based transactions. In line with the argument of Myers and Majluf (1984) these results suggest that in case of share based transactions acquirer shares may have been overvalued and acquirers may have bought targets at a discount.

In order to assess a potential change in abnormal returns over time we included a dummy variable for all transactions in the time period between 1998 and 2003. The results of the analyses suggest no change in abnormal returns over time as regression coefficients remain insignificant throughout all observed event windows.

Robustness of multivariate results: In order to assess the robustness of the multivariate regression analysis, we analyzed additional regression models on the BHARs for a 6 month, 18 month and 24 month holding period. Overall, the results are fairly consistent with our findings for a 12 month event window; both with regard to overall average abnormal performance as well as drivers of abnormal performance. While we do not individually report all analyzed regression models, a summary of the complete regression model over time is provided in Table A4.5.

4.5. Conclusion

The objective of this study was to analyze the long-term wealth effects of acquirers and rivals in the brewing industry. For this purpose, a sample of 66 horizontal M&A transactions involving brewing companies between 1998 and 2010 was identified and examined using a combination of two approaches: the buy-and-hold-return (BHAR) approach and the calendar time portfolio approach in combination with the Fama-French-3-Factor model (FF3F). Our results provide new insights into the perceived long-term success of M&A transactions in the brewing industry and its corresponding evaluation through capital markets.

First, our results indicate that acquirers in the brewing industry generally do not suffer from long-term value losses following M&A announcements. Our analysis suggests that overall abnormal performance is insignificant if not significantly positive. This finding is an outstanding attribute of the sector and stands in contrast to cross-industry studies by Loughran and Vijh (1997), Gregory (1997) and Rau and Vermaelen (1998) all of which provide consistent evidence of long-term value losses to acquirers. Therefore, it appears that capital markets value the above-average synergy potential of the sector.

Second, our study provides insights with regard to a number of value drivers of abnormal performance. Consistent with the findings of Conn et al. (2005), we determine a significant negative impact of cross-border acquisitions on long-term performance. At the same time, this negative impact is not confirmed for cross-border transactions involving targets in emerging market, suggesting that capital markets value the diversification strategies employed by many brewers. In contrast to the findings of Loughran and Vijh (1997), our study suggests a positive impact of share transactions on acquirer performance. On the other hand, we do not find any significant impact of transaction size, the target's public status or the transaction date.

Third, our results provide evidence that the "big four" cannot sustain the positive short-term announcement effects determined by Mehta and Schiereck (2011) in a companion study. Similarly, with regard to rival effects, we do not find evidence of significant abnormal long-term performance when missing out on a potential M&A opportunity, suggesting that the negative short-term rival effects as determined by Mehta and Schiereck (2011) are also not sustainable.

While our study addresses a number of important questions with regard to capital market effects of M&A in the brewing industry and the impact of potential determinant variables, the presented results also leave open issues and give rise to new research questions. As our study analyzes acquirer success solely from a capital market perspective, future research could additionally examine the impact of M&A on the operating performance of the involved companies. In general, given the expected continuation of the consolidation process in the sector, we believe that M&A in the brewing industry provide an interesting field for future research.

5. Study 3: Consolidation and Changes in the Risk Profile of the Brewing Industry [1]

5.1. Introduction

Empirical M&A research already discussed extensively the shareholder value effects of mergers and acquisitions (M&A). However, most of these studies implicitly assume that M&A have consequences for company wealth but no implications for company risk. This unproven assumption is critical given the fact that most analyses on wealth effects utilize the market model and estimate its parameters (i.e., alpha and beta) based on pre-event-estimation period returns, assuming the parameters to be stable during the specified event window. Very few researchers recognize that corporate events such as M&A might impact the risk characteristics of the involved firms. For instance, Conn (1985) argues that mergers in contrast to other corporate events such as earnings announcements are likely to impact a firm's operating and/or financial risk profile and thus also affect systematic risk. A measurement of wealth effects without considering potential changes in systematic risk is therefore incomplete and may lead to a biased description of M&A success. While a number of empirical studies have analyzed long-term changes in systematic risk as a result of M&A, the results are ambiguous with studies reporting positive beta changes (Langetieg, Haugen, and Wichern, 1980), negative beta adjustments (Mandelker, 1974; Lubatkin and O'Neill, 1987; Davidson, Garrison, and Henderson, 1987; Chatterjee and Lubatkin, 1990) as well as beta changes that are not significantly different from zero (Pettway and Yamada, 1986; Elgers and Clark, 1980). One explanation for this mixed evidence might be based on deviating risk changes in various industries which are not detected in cross-sectional analysis and which result in reported returns that depend on the composition of a given data set. To control for industry effects, we decide to focus on only one industry with an ongoing consolidation trend and a global competitive environment.

Towards the turn of the last century, many industries including the brewing industry have experienced a sharp increase in M&A activity. Consolidation has been and continues to be a major trend in the sector as multi-national breweries, in an attempt to offset the decline in beer volumes in mature markets (in particu-

---

[1] This paper is partly based on joint work between Malte Raudszus and Dirk Schiereck.

lar Western Europe), seek to expand their activities into new emerging markets. In contrast to many other sectors, the production, distribution, and marketing of beer is characterized by a relatively high fixed cost base, resulting in high levels of operational leverage (Earlam et al., 2010) providing larger brewers with material size advantages. Moreover, increased size has enabled brewers to build up a significant amount of market power (Slade, 2004), as larger brewers are able to negotiate favorable terms with their suppliers and benefit from greater bargaining power for negotiations with retail customers. Hence, it is not surprising that the global beer market today is dominated by large national/multinational companies rather than regional brewers. The four largest brewers Anheuser-Busch Inbev, Heineken, SABMiller and Carlsberg, the "big four", control about 50% of the global beer market (Jones, 2010). Despite rising market concentration, competition among the large brewing groups has remained fierce (Iwasaki, Seldon, and Tremblay, 2008). Going forward, industry experts predict that the consolidation process will continue (Fletcher, 2011). As sector debt levels are expected to decrease further, research analysts are certain that M&A will remain a major theme in the coming years as brewers will continue their quest for suitable M&A targets (Earlam et al., 2010).

In the light of these specific industry characteristics and the recent developments in the sector, the question remains open whether the M&A strategies employed by brewing companies are also reflected by capital markets in the form of abnormal risk shifts to acquiring brewing companies and their close rivals. More specifically, from a manager's perspective the question arises whether brewing companies, in addition to realizing potential revaluation of their equity, can alter systematic risk and whether it is persistently affected as a result of M&A announcements. Thus, with our analysis we aim to complement event study methodology, focusing on short-term abnormal returns. Comparable to event studies, a major concern for short-term risk measurement lies in dealing with event-induced variance. In the case of event studies, event-induced variance is best accounted for by applying parametric test-statistics that according to Harrington and Shrider (2007) use standard errors that are robust to cross-sectional variation in true abnormal returns. Advancing previous research, we analyze volatility behavior directly, measuring beta-risk dynamics around the event date. However, we separate event-induced variance from the event's persistent effect on the asset-to-market correlation coefficient. When forecasting beta around the event date we use actual daily market and asset standard deviations but the constant correlation coefficient form 41 trading days before the event. Subtracting actual from expected betas we shed light on short-term abnormal systematic risk changes.

Even though the global beer industry has gone through significant consolidation and seen a lot of M&A in recent years, empirical evidence remains scarce. Therefore, the aim of our study is also to fill this research gap and provide additional insights in M&A effects for investors and managers of beer companies. By this we contribute to the discussion of related topics around M&A, e.g., by Banal-Estañol and Seldeslachts (2011) who investigate merger failures and indentify the interaction between the pre- and post-merger processes as critical. An earlier study by Banal-Estañol and Ottaviani (2006) yet analyzed diversification effects around horizontal mergers, comparable to our brewery industry focus, finding a direct risk-sharing effect and an indirect strategic effect.

The remainder of this paper is structured as follows: Section 5.2 provides a brief overview of the relevant literature and outlines the derived hypotheses. Section 5.3 provides details on the applied methodology as well as the sample selection procedure. The following section 5.4 presents the empirical results and elaborates on the derived hypotheses. Finally, section 5.5 summarizes the findings and concludes.

5.2. Literature Review

5.2.1. Evidence on Capital Market Reactions around M&A Announcements

Even though most studies on wealth effects of M&A utilize the market model and estimate its parameters (alpha and beta) based on a pre-event estimation period, the stability of these parameters is questionable. The general concern is based on the perception that unlike corporate events such as earnings announcements, M&A may significantly impact the operating and/or financial risk profiles of acquirer and target companies (Kiymaz and Mukherjee, 2001; Conn, 1985). There are some studies investigating parameter shifts subsequent to M&A, but most of the studies focus on long-term risk changes. The evidence is mixed:

Joehnk and Nielsen (1974) analyze the immediate and long-term effects of M&A on beta for a sample of 44 acquiring firms. They categorize their sample into various subsamples including conglomerate and non-conglomerate transactions and consider relative differences in pre-event betas between acquirer and target companies. As one of six analyzed subsamples indicates a significant change in beta between pre- and post-event period, Joehnk and Nielsen conclude that overall their study shows weak evidence of the responsiveness of beta to M&A. They attribute the change in beta to be related to changes in operating and financial characteristics of the merging firms.

Based on a sample of 128 acquiring firms, Mandelker (1974) finds that the average beta increases in the pre-event period of -100 to -20 months by approximately 10% and subsequently decreases by 5% during the -20 to zero period and another 5% during the zero to +40 month period. Based on Mandelker's findings Conn (1985) concludes that cumulative abnormal returns are highly sensitive to the period chosen to estimate the parameters of the market model and therefore questions their validity.

Lubatkin and O'Neill (1987) analyze long-term changes in acquirer risk for 297 mergers, categorizing them based on the degree of the relatedness of the mergers. While the authors report significant increases in unsystematic risk for the entire sample, they provide evidence of a significant decline in systematic and total risk for related mergers. Consequently, the authors conclude that risk reduction may be a valid rationale for M&A. Davidson, Garrison, and Henderson (1987) come to a similar conclusion and attribute the risk reduction potential to merger synergies. More recently, Kiymaaz and Mukherjee (2001) confirm these results for cross-border transactions. The authors examine risk shifts for 112 transactions by US-based acquirers and report a significant decline in systematic risk in the post announcement period.

In contrast to the above-mentioned studies, that generally suggest a downward shift in the post-event beta, Langetieg, Haugen, and Wichern (1980) report a risk shift in the opposite direction. Based on a sample of 149 acquirers, the authors unexpectedly report a merger-induced increase in levels of systematic risk, total risk and unsystematic risk for the consolidated firm.

Hackbarth and Morellec (2008) argue that the sign of the risk change is dependent on the relative riskiness of acquirer and target. Using a real option model, the authors show that if the acquirer is more (less) risky than the target prior to the takeover, firm-level betas of acquirers increase (decrease) prior to the announcement and decrease (increase) in the subsequent 6 months.

On the other hand, there are also studies that report no evidence of post-event parameter changes. Based on an analysis of 59 transactions, Haugen and Langetieg (1975) conclude that there is only weak and thus no sufficient evidence that M&A lead to parameter changes. Similarly, Dodd (1980), in a study analyzing wealth effects of M&A, does not find differences in CARs using various estimation periods. While the author does not provide details on the conducted sensitivity tests, Connell and Conn (1993) argue that these results stand in contrast to the findings of Mandelker (1974). After analyzing a sample of 337 acquiring firms Elgers and Clark (1980) find that the systematic risk of the buyer portfolio approximates the market beta and is essentially unchanged between the pre- and post-merger period. Likewise, Pettway and Yamada (1986) in a study of Japanese M&A transactions do not find sufficient evidence to explain an impact of M&A on systematic or unsystematic risk.

Amihud, DeLong, and Saunders (2002) come to a similar conclusion for cross-border bank mergers and find that systematic and total acquirer risk neither increase nor decrease as a result of an M&A transaction.

Empirical research also investigates the impact of M&A on rival firms. However, all of the reviewed studies focus on wealth effects of rival companies and do not specifically analyze implications of M&A on rival risk. With regard to wealth effects, researchers' findings show positive as well as negative effects on rivals: On the one hand, rival companies may benefit from the transaction announcement as a result of a positive signaling effect regarding industry attractiveness and potential future takeover activity (Eckbo, 1983; Song and Walkling, 2000). Moreover, a merger in the industry decreases the number of competitors and thus increases the likelihood of collusion, which may lead to greater monopoly rents to rival firms (Eckbo, 1983; Shahrur, 2005). On the other hand, rival firms may be impacted by negative competitive effects due to more intense competition in the industry from a new, more-efficient combined firm (Eckbo, 1983). Overall, the positive effects outweigh the negative competitive effects: For example, Eckbo (1985) reports positive announcement effects to rivals for horizontal transactions. Similar results are found by Song and Walkling (2000) in a study including horizontal and non-horizontal transactions. These results are confirmed in more recent studies by Clougherty and Duso (2009), Fee and Thomas (2004), and Shahrur (2005).

As mentioned above, empirical research on M&A in the brewing industry remains scarce and primarily focuses on the U.S. brewing industry. The specific topics addressed in the studies focus on technological change in the sector (Kerkvliet et al., 1998), the sector's tendencies towards concentration (Lynk, 1985; Adams, 2006), the determinants and motives for horizontal M&A (Tremblay and Tremblay, 1988), as well as competition in the industry (Horowitz and Horowitz, 1968). More recently, Ebneth and Theuvsen (2007) analyze the short-term value effects of M&A to acquirers using event study methodology. Based on a sample of 29 cross-border transactions involving European acquirers from 2000-2005, they find insignificant positive acquirer returns. In a companion paper, Mehta and Schiereck (2011) challenge these results and report significant positive abnormal returns to acquirers. Moreover, they find a significant negative effect to close rivals missing out on a potential M&A opportunity.

5.2.2. Contribution to Literature and Research Focus

With this paper we aim to contribute to existing literature in two ways. First, we provide a comprehensive analysis of short-term risk effects of M&A to acquirers in the brewing industry using a global dataset that in addition to cross-border transactions also includes domestic transactions. This enables us to specifically build and investigate relevant subsamples and determine key drivers of systematic risk adjustments. Second, we specifically analyze abnormal risk shifts of M&A transactions by the "big four" as well as the competitive effects of missing out on a transaction.

As pointed out above, previous studies report mixed results. Given the particularities of the brewing industry and its development over the last few years, we expect M&A in the sector to be a feasible measure to realize synergy and efficiency gains. However, given the fairly concentrated market structure, where it is becoming increasingly hard to find suitable targets, we believe that overall horizontal diversification potential is limited. While certain transactions might have an unexpected risk-reducing effect, for the total sample we do not expect to find a significant abnormal change in systematic risk.

In order to determine potential drivers of systematic risk adjustments, a number of hypotheses concerning the impact of deal, acquirer, and target characteristics are formulated.

Geographical scope: Connell and Conn (1993) investigate abnormal returns for 73 mergers between US and UK firms using three different estimation periods (pre-event, post-event, and pooled). The authors find that abnormal returns are highly sensitive to the estimation period chosen for the market model. More specifically, they report lower abnormal returns when a post-merger estimation period is employed and attribute 85% of the difference to decline in alphas. On the other hand, Fatemi (1984), while not specifically investigating cross-border mergers, finds a negative relationship between the international involvement of a bidder and the post-merger beta. As pointed out above, Kiymaz and Mukherjee (2001) confirm these results and report a significant decline in systematic risk in the post event period for cross-border transactions by US acquirers. The authors conclude that cross-country diversification provides risk reduction and thus value-enhancing opportunities, which cannot be replicated by investors using portfolio diversification.

In case of the brewing industry, cross-border transactions have been of particular relevance as many brewers have looked to diversify into foreign markets given the decline in beer volumes in many mature markets. In particular, brewers have looked at acquisition opportunities in emerging markets such as Asia, Eastern Europe, and Latin America, which are still relatively fragmented and offer growth opportunities for beer volumes. From a risk perspective, we assume cross-border M&A to be a feasible measure for brewers to diversify their earnings streams and thus expect to find a significant risk-reducing effect of cross-border M&A on acquirers, in particular when acquired targets are based in emerging market economies. Moreover, we expect to find significant differences between cross-border and domestic transactions.

Transaction value and relative transaction size: The brewing industry has seen a significant increase in transaction values in recent years. Given the particular characteristics of the sector, increased company size provides brewers with a material competitive advantage. Consequently, it can be argued that the acquisition of big targets will significantly contribute to the success of a transaction due to greater potential for revenue and cost synergies. From a risk perspective, a larger target may provide a greater diversification opportunity. On the other hand, the integration of larger targets is likely to be more difficult than for small targets (Hawawini and Swary, 1990) and thus potentially also more risky. In addition to target size, the size of the acquirer may also influence the success of a transaction. Moeller, Schlingemann, and Stulz (2003) argue that managers of larger companies are more likely to overestimate their own abilities. Due to the fact that larger companies may benefit from bigger cash reserves and also might need to face fewer hurdles in the execution of M&A transactions, the authors argue that managers of larger companies are more likely to engage in M&A transactions that are not always beneficial to the company.

Transaction date: Over the last few years, the brewing industry has experienced a sharp increase in M&A activity and seen strong industry consolidation. As a consequence, the competitive landscape has materially changed as the "big four" today control more than 50% of global beer volumes. While the sector is expected to further consolidate in the future, it is becoming increasingly difficult to find suitable targets (Gibbs, Webb, and Dhillon, 2010). We assume the changes in market structures and concentration to have an impact on acquirer returns and expect to find significant differences in risk effects over time.

As already mentioned above, the global beer market is dominated by Anheuser-Busch Inbev, SABMiller, Heineken, and Carlsberg, the "big four". In a companion paper, analyzing M&A announcement effects, Mehta and Schiereck (2011) specifically analyze rival effects among the "big four" and find significantly positive abnormal returns if one of the "big four" announces an M&A acquisition, while the remaining three rivals, missing out on an M&A opportunity experience negative abnormal returns. Similarly, we expect to find a stabilizing effect if one of the "big four" engages in an M&A transaction and thus predict a decline in systematic risk. Given the strong competition among the "big four" we also expect to find an impact on the risk levels of the remaining three rivals as they are put into a disadvantageous position by missing out on a diversification and growth opportunity in a fairly concentrated market.

5.3. Data Selection and Research Methodology

5.3.1. Data Selection

The sample of mergers and acquisitions for the risk event study is drawn from the Securities Data Corporation (SDC)/Thomson One Banker Deals database and the Merger Market M&A database. It includes all world-wide M&A events announced between January 1st, 1991, and December 31st, 2010. The total number of M&A deals is reduced to yield only those transactions meeting the following criteria:

i.   At the time of the transaction, acquirer and target companies both had active operations in the brewing industry.

ii. The acquiring company has been publicly listed for at least 250 days prior to the announcement of the transaction.

iii. The total transaction value accumulates to at least USD 50 million.

iv. The completion of the transaction leads to a change of control in the target; prior to the announcement of the transaction the bidder holds less than 50% in the target company, following the transaction the bidder obtains a controlling stake in the target company.

v.   The transaction has been successfully completed.

In addition, the transactions were validated by a press research using the Factiva database as well as company websites in order to ensure that all transactions are horizontal and the announcement dates provided by the databases are correct. Moreover, acquirers with multiple transactions on the same day as well as acquirers with limited available trading data were removed from the dataset. The described selection criteria result in a final sample of 75 transactions. The frequency distribution of the transactions over time is provided in Table 5.1. While the number of transactions is spread fairly even from 1998 onwards, the average transaction size varies strongly from USD 50 million to USD 11,973 million due to a number of high-profile transactions such as InBev's acquisition of Anheuser-Busch (USD 52 billion), Heineken and Carlsberg's takeover of S&N (USD 15 billion) and Heineken's recent acquisition of Femsa Cerveza

(USD 5.7 billion). In terms of geography more than 75% of the transactions in-volve acquirers that are based in Europe.

Table 5.1: Sample Overview: Descriptive Statistics

Year	Deals	(%)	Avg. Trans. Val. (USD mil.)	Trans. Val. (USD mil.)	Acquirer Region - Number of Deals			
					Europe	Americas	Asia	RoW
2010	1	1.3	5,700	5,700	1			
2009	4	5.3	737	2,946	2		2	
2008	6	8.0	11,973	71,835	6			
2007	5	6.7	405	2,025	1	2	1	1
2006	6	8.0	332	1,991	6			
2005	8	10.7	785	6,281	8			
2004	9	12.0	875	7,873	6	2	1	
2003	6	8.0	582	3,489	6			
2002	6	8.0	1,526	9,157	5	1		
2001	4	5.3	599	2,396	3	1		
2000	6	8.0	514	3,086	5		1	
1999	4	5.3	495	1,981	3	1		
1998	4	5.3	268	1,071	4			
1997	0	0.0	0	0				
1996	1	1.3	65	65			1	
1995	1	1.3	991	991	1			
1994	1	1.3	134	134		1		
1993	1	1.3	155	155				1
1992	1	1.3	79	79	1			
1991	1	5.8	50	50	1			
Sum	75	100.0	1,617.4	121,305.0	59	9	5	2

Sample includes all M&A transactions between 1998 and 2010 as specified above. Table 5.1 shows the frequen-cy distribution of M&A transactions between 1998 and 2010 with total and average transaction values in USD mil. Additionally, details on acquirer region are provided.

The relevant daily stock prices, market capitalizations, and local market in-dices for acquirers were downloaded from the Thomson Datastream database. Acquirer returns are calculated using the Datastream Total Return Index, which adjusts the closing share prices for dividend payments as well share issuances or repurchases.

5.3.2. Research Methodology

In order to determine and measure risk shifts we apply an event study methodology that has been advanced to measure short-term shifts in beta that exceed or fall below expectations. The mathematical intuition is highly comparable to the standard event study. Our risk event study is composed of three basic steps: first, measuring daily acquirer betas, second, estimating expected betas from before M&A announcements, and third, subtracting measured betas from expected betas to obtain abnormal beta shifts, AB. Cumulating ABs within event windows and determining the arithmetic mean across events finally returns the cumulative average abnormal beta shifts, CAAB. Intuitively obvious to scholars of financial economics measuring time-variant beta breaches the basic assumption of the CAPM stating that beta is stable over time. But, financial economists like Jagannathan and Wang (1996), Collins, Ledolter, and Rayburrn (1987), Bos and Newbold (1984), Sunder (1980), and Fabozzi and Francis (1978) have argued against the assumption of beta stability and showed that beta is actually stochastic and thus conditional.

To determine time-varying volatility and beta measures we apply the generalized autoregressive conditional heteroscedasticity (GARCH) approach. It is one of the most commonly applied techniques by researchers to observe changes in risk levels. The methodology was first introduced by Engle (1982), refined by Bollerslev (1986) and further advanced to a Multivariate GARCH (M-GARCH) by Bollerslev (1990). See Giannopoulos (1995) for an empirical implementation of GARCH models where he finds evidence that such volatility modeling techniques cannot only be used to measure total risk but also systematic and unsystematic components. Brooks, Faff, and McKenzie (1998) investigate the forecasting accuracy of three techniques regarding time-variant beta: a M-GARCH, a time-varying beta market model suggested by Schwert and Seguin (1990), and the Kalman filter for a sample of returns on Australian industry portfolios. Furthermore, Faff, Hillier, and Hillier (2000) applied an M-GARCH approach to model time-varying beta in their study discussing alternative beta risk modeling techniques. They define an additional assumption of a constant correlation coefficient between the returns of an asset and respective market returns. The behavior of this correlation coefficient is core to our risk event study determining CAABs.

We apply a simple GARCH (1,1) model. As shown by Equation (5.1) it determines a specific asset's time-varying volatility in terms of the variance rate, σ_{it}^2, as a weighted sum of a long-term average variance, Φi, the previous day squared return, $u_{i,t-1}^2$, and the previous day variance, $\sigma_{i,t-1}^2$. The weighting factors, γ1, γ2, and γ3 are determined applying the iterative Maximum Likelihood technique as discussed by Hull (2006).

$$\sigma_{it}^2 = \gamma_1 \Phi_i + \gamma_2 u_{i,t-1}^2 + \gamma_3 \sigma_{i,t-1}^2 \qquad (5.1)$$

Obtaining beta measures as next step, we apply the M-GARCH model that observes the actual betas as product of a constant correlation coefficient, ρim, between the specific acquirer returns and the respective market returns, and the time-varying acquirer standard deviation, σit, divided by the time-varying market standard deviation, σmt.

$$\beta_{it} = \frac{\rho_{im} \sigma_{it}}{\sigma_{mt}} \qquad (5.2)$$

Finishing the above elucidated empirical groundwork, we have the following data at hand: Daily volatility measures for all brewery industry acquirers and respective market indices; daily correlation coefficients for all acquirers with their corresponding market indices based on a one year lag structure; and daily beta measures for all acquirers.

Equivalent to beta for the standard return event study we utilize the correlation coefficient 41 trading days prior to the event (i.e., 21 trading days or one month before the start of the maximum event window [-20;+20]) for our systematic risk estimates, $E(\beta_{it})$. The correlation coefficient is determined over a 250 trading day period comprising one calendar year. Acquirer and market volatility measures remain time-variant as the market return does in the standard event study. The abnormal beta shift, AB, is then determined as difference between expected and measured beta from up to ten days before to up to ten days after the event.

$$AB_{it} = E(\beta_{it}) - \beta_{it} = \frac{\rho_{im} \sigma_{it}}{\sigma_{mt}} - \beta_{it} \qquad (5.3)$$

Cumulating ABs along the differently sized event windows we obtain cumulative abnormal beta shifts, CAB. The arithmetic mean of CABs across the number of events, n, per analysis returns the cumulative average abnormal beta shifts, CAAB. The Kolmogorov-Smirnov test is used to determine if the CABs are normally distributed and hence to decide whether to test the CAAB means and medians for difference to zero with parametric t-tests or non-parametric Wilcoxon signed rank tests.

Table 5.2 provides an overview on average market and brewery risk measures such as variance, correlation coefficient, and beta. Worth mentioning is the by far higher average brewery variance than market variance. In addition, more specific brewery risk adjustments with respect to the event-window effects of variance and correlation on beta can be directly observed. The correlation coefficient drops from 41 days prior the event to the event date, whereas the variance increases in parallel. In sum, beta rises and this upward trend is significant at the 10% level, even though shifts in correlation and variance are not statistically significant. The increase in brewery variance towards the event date and thereafter is what we refer to as event-induced variance. On the other hand, we define an event's effect on the correlation coefficient as a more persistent effect on beta and consequently focus our analysis thereon.

Table 5.2: Total Sample Risk Characteristics

		Correlation	Variance	Beta		N
1997-2010	market ind.	-	0.00028	-		16
average	breweries	0.423	0.00045	0.563		26
	-40	0.454	0.00036	0.634		75
Event window	Event date	0.449	0.00038	0.666		
	20	0.450	0.00041	0.665		
	-41/0	-0.006	0.00003	0.033	*	75
		-1.0	*0.9*	*1.6*		
Adjustments	0/20	0.001	0.00003	-0.001		
(t-stat.)		*0.2*	*0.6*	*-0.1*		
	-41/20	-0.006	0.00005	0.031	*	
		-0.8	*1.1*	*1.6*		

***, **, and * indicate statistical significance at the 1%, 5%, and 10% levels, respectively. t-test: std. parametric mean test. Contents of "Event-window" and "Adjustments" sections specify brewery risk measures in relation to respective market indices.

5.4. Empirical Results

5.4.1. Acquirer Short-term Systematic Risk Changes

Figure 5.1 reports the short-term abnormal risk effects of M&A transactions on the total sample of acquirers in the brewing industry. The results show that for the total sample all daily average abnormal beta changes (AAR) are negative. We observe a strong increase in abnormal daily beta descents beginning from the event date though decreasing towards the end of the event window. Nevertheless, as the results in Table A5.1 in the appendix show, the observable daily risk adjustments hardly lead to any significant CAABs. These findings are in line with our predictions. In the following, we will analyze relevant subsamples to gain further insight on a potential relationship between M&A announcements and beta changes.

Figure 5.1: Daily Acquirer AABs throughout the Event Window

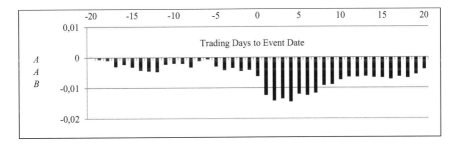

Geographical scope: In order to assess the impact of geographical diversification on acquirer beta changes we analyze domestic transactions, cross-border transactions and emerging market transactions (i.e., transactions where the acquired targets are based in emerging market economies). The results are provided below, Table 5.3 and in the appendix, Table A5.2. In line with our predictions we find a significant decrease in beta for cross-border transactions across almost all examined event windows. Acquirers experience a significant abnormal beta reduction of 0.55 and 0.29 for the [-20; 20] and [-10; 10] event window, respectively. These results suggest that cross-border transactions seem to have a stabilizing effect on acquirer systematic risk. We find similar results for cross-border transactions with targets in emerging market economies (Asia ex Japan, Latin America, and Eastern Europe). CAABs are significant for all examined event windows. For the [-20; 20] and [-10; 10] event windows we find negative beta changes of -0.51 and -0.23, respectively, both significant at the 5% level. In contrast, we find no impact on systematic risk for domestic transactions (see Table A5.2). While these results suggest a relationship between the geographical scope of a transaction and systematic risk, we will challenge these results with a multivariate regression analysis.

Table 5.3: Acquirer CAAB by Geographical Scope

CAAB	[-x; +y]	Mean	Median	KS-test	p	Wx-test	p	t-test	p	n
Cross-Border M&A	0 0	-0.01	-0.01	0.11	0.53	-1.50	0.13	-0.02	0.28	51
	1 1	-0.03	-0.03 *	0.11	0.56	-1.80	0.07	-0.07	0.16	51
	2 2	-0.06	-0.05 *	0.10	0.63	-1.76	0.08	-0.13	0.13	51
	5 5	-0.12	-0.15 **	0.10	0.59	-1.92	0.05	-0.28	0.12	51
	10 10	-0.23	-0.29 **	0.11	0.50	-2.05	0.04	-0.53	0.13	51
	20 20	-0.42	-0.55 **	0.11	0.54	-2.09	0.04	-0.96	0.12	51
	0 1	-0.02	-0.02 *	0.13	0.36	-1.82	0.07	-0.05	0.12	51
	0 2	-0.04 *	-0.03 *	0.12	0.43	-1.87	0.06	-0.09	0.09	51
	0 5	-0.10 *	-0.07 **	0.10	0.62	-2.05	0.04	-0.20	0.08	51
	0 10	-0.19 *	-0.16 **	0.10	0.63	-2.02	0.04	-0.40	0.08	51
	0 20	-0.34	-0.27 **	0.09	0.78	-1.92	0.05	-0.76	0.11	51
Emerging Market M&A	0 0	-0.01	-0.01 *	0.16	0.31	-1.70	0.09	-0.03	0.17	34
	1 1	-0.04	-0.03 **	0.17	0.24	-1.97	0.05	-0.10	0.11	34
	2 2	-0.08 *	-0.05 **	0.17	0.25	-2.06	0.04	-0.17	0.09	34
	5 5	-0.17 *	-0.08 **	0.16	0.33	-2.15	0.03	-0.36	0.08	34
	10 10	-0.32 *	-0.23 **	0.17	0.22	-2.27	0.02	-0.67	0.07	34
	20 20	-0.58 **	-0.51 **	0.14	0.51	-2.40	0.02	-1.17	0.05	34
	0 1	-0.04 *	-0.02 **	0.17	0.27	-1.94	0.05	-0.08	0.08	34
	0 2	-0.06 *	-0.02 **	0.19	0.17	-2.21	0.03	-0.13	0.07	34
	0 5	-0.13 *	-0.07 **	0.19	0.13	-2.38	0.02	-0.27	0.06	34
	0 10	-0.26 **	-0.16 **	0.19	0.13	-2.37	0.02	-0.50	0.04	34
	0 20	-0.48 **	-0.28 **	0.19	0.14	-2.32	0.02	-0.93	0.04	34

***, **, and * indicate statistical significance at the 1%, 5%, and 10% levels, respectively. *CAAB*: cumulative average abnormal beta shift, KS test: Kolmogorov-Smirnov test - p-value ≤ 0.05 indicates normal distribution at the 5% significance level, Wx-test: Wilcoxon signed rank test as non-parametric median test, t-test: std. parametric mean test.

Transaction value and relative transaction size: Table 5.4 presents the CAABs for the 30 largest transactions. In line with our expectations we do not find any abnormal beta changes for the 30 smallest transactions and hence leave them out. On the other hand, we report significant negative abnormal beta changes for large transactions across all analyzed event-windows. For the [-20; 20] and [-10; 10] event windows we find negative beta changes of -0.77 and -0.42, respectively, both significant at the 5% level. The results are in line with our predictions and suggest a potentially negative relationship between transaction value and beta. Again, we will challenge the risk event study results in the multivariate analysis.

Table 5.4: Acquirer CAAB of 30 Largest M&A by Transaction Value

CAAB [-x; +y]		Mean		Median		KS-test	p	Wx-test	p	t-test	p	n
0	0	-0.01		-0.01	*	0.15	0.45	-1.70	0.09	-0.03	0.12	30
1	1	-0.06	**	-0.06	**	0.15	0.46	-2.09	0.04	-0.12	0.04	30
2	2	-0.11	**	-0.10	**	0.14	0.58	-2.13	0.03	-0.21	0.03	30
5	5	-0.24	**	-0.21	**	0.12	0.77	-2.27	0.02	-0.45	0.03	30
10	10	-0.43	**	-0.42	**	0.11	0.81	-2.27	0.02	-0.83	0.04	30
20	20	-0.73	**	-0.77	**	0.11	0.78	-2.23	0.03	-1.47	0.05	30
0	1	-0.04	**	-0.04	**	0.18	0.25	-2.07	0.04	-0.09	0.04	30
0	2	-0.08	**	-0.06	**	0.17	0.34	-2.13	0.03	-0.15	0.03	30
0	5	-0.17	**	-0.12	**	0.14	0.58	-2.25	0.02	-0.33	0.03	30
0	10	-0.32	**	-0.22	**	0.15	0.45	-2.25	0.02	-0.62	0.04	30
0	20	-0.56	*	-0.55	**	0.14	0.57	-2.07	0.04	-1.17	0.07	30

***, **, and * indicate statistical significance at the 1%, 5%, and 10% levels, respectively. CAAB: cumulative average abnormal beta shift, KS test: Kolmogorov-Smirnov test - p-value ≤ 0.05 indicates normal distribution at the 5% significance level, Wx-test: Wilcoxon signed rank test as non-parametric median test, t-test: std. parametric mean test. The top 30 M&As by transaction value include transactions between USD 400 mil. and 53 bil.

Table A5.3 reports the CAABs for the 30 smallest and 30 largest transactions by relative transaction size, which is determined as transaction value divided by acquirer market value. While we find significant beta shifts in some event windows, overall there does not seem to be enough evidence to suggest a potential relationship between relative transaction size and systematic risk. Nonetheless, the relative size of transactions will be considered as part of the multivariate regression analysis.

Transaction date: As pointed out in the literature review, the structure of the brewing industry has materially changed in the course of the last years. Table A5.4 presents changes in systematic risk for transactions from 1991-2003 and 2004-2010. We find insignificant negative beta changes for both subsamples. These results are similar to the results for the entire sample and do not suggest an impact of transaction date on systematic risk.

As already mentioned in section 5.2, the "big four" dominate the global brewing industry. Table 5.5 presents the CAABs for transactions by the "big four". We find significant negative abnormal beta changes across all examined event windows. For the [-10; 10] and [-20; 20] event windows we find negative CAABs of -0.35 and -0.62, respectively. These results are in line with our predictions. It should be noted that all "big four" transactions in the sample were cross-border transactions. Overall, it appears justified to conclude that the "big four" have been able to reduce systematic risk by engaging in M&A.

Table 5.5: Acquirer CAAB of "big four"

CAAB [-x; +y]	Mean	Median		KS-test	p	Wx-test	p	t-test	p	n
0 0	-0.01	-0.02	*	0.15	0.36	-1.78	0.08	-0.03	0.31	36
1 1	-0.04	-0.05	**	0.13	0.52	-2.06	0.04	-0.09	0.17	36
2 2	-0.06	-0.08	**	0.14	0.41	-2.04	0.04	-0.15	0.14	36
5 5	-0.15	-0.19	**	0.13	0.53	-2.15	0.03	-0.34	0.13	36
10 10	-0.28	-0.35	**	0.15	0.33	-2.18	0.03	-0.65	0.13	36
20 20	-0.51	-0.62	**	0.13	0.54	-2.22	0.03	-1.16	0.12	36
0 1	-0.03	-0.04	**	0.14	0.43	-2.07	0.04	-0.07	0.13	36
0 2	-0.05 *	-0.05	**	0.15	0.37	-2.17	0.03	-0.11	0.10	36
0 5	-0.12 *	-0.09	**	0.15	0.35	-2.29	0.02	-0.25	0.09	36
0 10	-0.23 *	-0.21	**	0.16	0.30	-2.23	0.03	-0.49	0.08	36
0 20	-0.42 *	-0.53	**	0.12	0.61	-2.18	0.03	-0.94	0.10	36

***, **, and * indicate statistical significance at the 1%, 5%, and 10% levels, respectively. *CAAB*: cumulative average abnormal beta shift, KS test: Kolmogorov-Smirnov test - p-value ≤ 0.05 indicates normal distribution at the 5% significance level, Wx-test: Wilcoxon signed rank test as non-parametric median test, t-test: std. parametric mean test.

5.4.2. Rival Short-term Systematic Risk Changes

Table 5.6 presents the rival effects among the "big four" i.e., the abnormal beta changes for those three companies of the "big four" missing out on a potential M&A opportunity when one of their close rivals announces an M&A transaction. In line with our predictions, we find a significant impact on systematic risk of rival companies. While average and median cumulative beta changes are positive across all analyzed event windows, we find significant positive CAABs for a number of the analyzed event windows with significance at the 5% and 10% level. In order to detect potential drivers of abnormal rival beta changes we specifically analyze various subsamples similar to our acquirer analysis (see Tables A5.5, A5.6).

Table 5.6: Rival CAAB among "big four"

CAAB [-x; +y]	Mean	Median		KS-test	p	Wx-test	p	t-test	p	N
0 0	0.00	0.01	**	0.14	0.03	-2.03	0.04	-0.01	0.49	96
1 1	0.01	0.02	**	0.15	0.02	-1.93	0.05	-0.02	0.51	96
2 2	0.02	0.03	*	0.16	0.01	-1.86	0.06	-0.04	0.56	96
5 5	0.03	0.03	'	0.16	0.01	-1.70	0.09	-0.09	0.65	96
10 10	0.03	0.09		0.17	0.01	-1.46	0.14	-0.17	0.74	96
20 20	0.07	0.15		0.16	0.02	-1.47	0.14	-0.30	0.71	96
0 1	0.01	0.01	**	0.15	0.03	-2.01	0.04	-0.01	0.46	96
0 2	0.01	0.02	*	0.14	0.04	-1.82	0.07	-0.02	0.54	96
0 5	0.01	0.04		0.14	0.03	-1.53	0.12	-0.05	0.66	96
0 10	0.02	0.05		0.13	0.07	-1.25	0.21	-0.10	0.80	96
0 20	0.03	0.04		0.13	0.08	-0.98	0.33	-0.20	0.79	96

***, **, and * indicate statistical significance at the 1%, 5%, and 10% levels, respectively. *CAAB*: cumulative average abnormal beta shift, KS test: Kolmogorov-Smirnov test - p-value ≤ 0.05 indicates normal distribution at the 5% significance level, Wx-test: Wilcoxon signed rank test as non-parametric median test, t-test: std. parametric mean test.

Geographical scope: We find significant positive abnormal beta shifts for rivals if a transaction is announced in a mature market as shown in Table 5.7. Beta changes are significant at the 5% or 10% level across almost all analyzed event windows. In contrast to most emerging market economies where beer markets are generally relatively fragmented, most mature beer markets are rather concentrated and dominated by few market players. Consequently, the consolidation potential in mature markets is generally smaller, hence providing brewers with fewer diversification opportunities. Thus, a possible explanation for the increase in systematic risk for rivals may be the fact that they missed out on a potential consolidation opportunity in a fairly concentrated market while at the same time a relevant competitor strengthened its position in the same market. We will verify these results in the upcoming multivariate regression analysis. Since we do not find any abnormal risk changes for acquisitions regarding emerging markets, we do not report respective statistics.

Table 5.7: Rival CAAB among "big four" in Mature Markets

CAAB [-x; +y]	Mean		Median		KS-test	p	Wx-test	p	t-test	p	n
0 0	0.02		0.02	*	0.14	0.46	-1.85	0.06	0.00	0.06	34
1 1	0.05	*	0.05	**	0.12	0.65	-1.96	0.05	0.00	0.06	34
2 2	0.08	**	0.08	*	0.12	0.66	-1.85	0.06	0.00	0.05	34
5 5	0.16	**	0.13	*	0.12	0.68	-1.87	0.06	0.00	0.05	34
10 10	0.27	*	0.22	*	0.14	0.50	-1.75	0.08	-0.02	0.07	34
20 20	0.45	*	0.43	*	0.15	0.43	-1.75	0.08	-0.03	0.07	34
0 1	0.03	**	0.04	**	0.14	0.48	-2.01	0.04	0.00	0.05	34
0 2	0.05	**	0.06	**	0.14	0.52	-2.03	0.04	0.00	0.05	34
0 5	0.09	*	0.09	*	0.12	0.69	-1.68	0.09	-0.01	0.09	34
0 10	0.13		0.14		0.12	0.62	-1.55	0.12	-0.05	0.14	34
0 20	0.22		0.22		0.09	0.89	-1.38	0.17	-0.12	0.19	34

***, **, and * indicate statistical significance at the 1%, 5%, and 10% levels, respectively. *CAAB*: cumulative average abnormal beta shift, KS test: Kolmogorov-Smirnov test - p-value ≤ 0.05 indicates normal distribution at the 5% significance level, Wx-test: Wilcoxon signed rank test as non-parametric median test, t-test: std. parametric mean test.

Transaction value and relative transaction size: While we find weak evidence for significant positive risk shifts for rivals for the smallest 30 transactions by transaction value (see Table A5.5), we do not believe that the risk event study provides sufficient evidence to explain a potential relationship between transaction value or relative transaction size and rival risk shifts.

Transaction date: We find significant abnormal positive risk shifts for rivals for transactions between 1991 and 2003 (see Table A5.6). The results are significant at the 5% or 10% level across almost all analyzed event windows. On the other hand, we do not find abnormal risk shifts for transactions between 2004 and 2010. The results are in line with our analysis on the geographical scope of transactions, as the majority of transactions between 1991 and 2003 (>54% of transactions) involved targets in mature markets, while most of the transactions between 2004 and 2010 (>75% of transactions) involved targets in comparably fragmented emerging market economies. Another possible explanation for the results could be seen in the emergence of the "big four" as well-diversified global players over time. Consequently, it could be argued that rivals were more affected by competitors' M&A activities while they were still comparably small and not as well diversified. Again the results will be verified in the multivariate regression analysis.

5.4.3. Summary of Acquirer and Rival Risk Changes

Figure 5.2: Cumulative Acquirer AABs throughout the Event Window

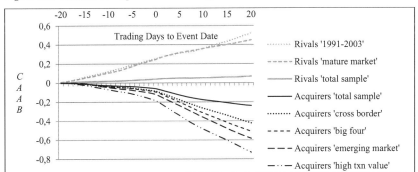

The risk event study analysis of acquirers and rivals has shown various abnormal risk changes around the announcement of M&A transactions. Thereby, two tendencies can be identified: First, on average, acquirers could realize significant abnormal descents in systematic risk around the event date if certain transaction characteristics such as cross-border M&A, target in emerging market, or market leading acquirer (big four) were observed. Second, rivals showed significant abnormal increases in systematic risk throughout the event window – especially, if missing out on targets in mature markets where respective M&A largely took place during the 1991 to 2003 time period. Figure 5.2 illustrates the accumulation of daily average abnormal beta changes from 20 trading days before the transaction [-20;-20] to the end of the entire event window [-20;+20]. The CAABs at the end of the event window can also be found in the respective tables where the significance levels are indicated.

Overall, the risk event study results support our insights gained from the literature review. The abnormal beta descent for the total acquirer sample is not significant, but the risk diversifying effect of cross-border M&A and by that especially the acquisition of targets in growing emerging markets confirms our expectations. Furthermore, large transactions emerge as a major driver for decreasing beta risk, supporting the argument for extended cost synergy potential in the trade-off to an increased risk of a successful integration of a larger target company. The risk-reducing effect of "big four" transactions, which coincide significantly with cross-border, emerging market, and large transaction value M&A, serves as evidence for their shareholder value enhancing growth strategy.

93

Missing out on transactions results in a significant abnormal beta risk increase for rival breweries which is further enhanced if M&A take place in mature markets. These findings also confirm our expectations of rivals being put into a disadvantageous position having reduced their diversification potential in a fairly contracted market.

5.4.4. Determinants of Abnormal Risk Changes

In order to obtain further insights on potential drivers of systematic risk changes as yet indicated by the risk event study, two cross-sectional regressions are performed on the cumulative abnormal beta changes to acquirers and rivals.

Our first multivariate regression analysis includes up to six variables to represent the parameters, which have already been analyzed individually in the risk event study subsamples. Equation 5.4 shows the respective model with the acquirer CAB of the $[0;+5]$ event window as dependent variable. The independent variables are discussed in the following paragraphs.

$$CAB_{acqu.} = \alpha + \delta_1 * CB + \delta_2 * EM + \delta_3 * B4 + \delta_4 * TV + \delta_5 * RS + \delta_6 * Yr$$

(5.4)

Geographical scope: The results of the risk event study suggested that the announcement of cross-border transactions, when compared to domestic transactions, might lead to a significant reduction in systematic risk. At the same time, the results suggested a significant decline in systematic risk if targets were based in emerging market economies (Latin America, Eastern Europe, and Asia excluding Japan). Both effects are included in the regression model using the dummy variables CB for "cross-border M&A" and EM for "target emerging market".

"big four" transaction: Risk event study subsample results also showed significant negative cumulative abnormal beta shifts for transactions announced by the "big four". We test these results by including a dummy variable B4 for transactions by the "big four".

Transaction value and relative transaction size: In line with our predictions, the risk event study suggests a potential negative relationship between transaction value and beta. On the other hand, we did not find enough evidence suggesting an impact of relative transaction size (transaction value / acquirer market value) on systematic risk. In order to verify these results, transaction value, TV, and relative transaction size, RS, are included as variables in the regression model.

Transaction date: As pointed out above, the structures of the global beer market have materially changed in recent years. In the risk event study subsample analysis we did not find significant beta changes for any time period analyzed. Nonetheless, we include the year, Yr, of the M&A announcement in the regression model to again scrutinize this potential driver.

Table 5.8 presents the results of four regression models (M1-M4) on the CABs for the [0;5] event window. This specific event window has been chosen because it is the only one that showed a significant abnormal average beta change for acquirers within the risk event study. If we may expect any significant result in the respective multivariate analysis, it will most definitely be observable within the [0;5] event window. The regression models differ with regard to the independent variables included. Overall, none of the models is statistically significant (based on F-statistics). The explanatory power is low with an adjusted R-squared ranging between -3% and 2%. Autocorrelation issues can be ruled out due to Durbin-Watson-statistics near two in both cases. The constant, alpha, is insignificant for all four models. In general, these findings are in line with our expectations and confirm the largely insignificant beta changes determined for the total sample in the risk event study.

Table 5.8: Multivariate Regressions on Acquirer CABs

Dependent variable		CAB [0;5]			
Model		M1	M2	M3	M4
Intercept	α	2.49	0.02	-0.02	-0.05
		0.1	*0.4*	*-0.4*	*-1.1*
Independent variables					
Cross-border M&A	δ_1	-0.03			
		-0.3			
Target emerging market	δ_2	-0.13	-0.14 *	-0.11	
		-1.5	*-1.6*	*-1.2*	
Acquirer size (Big Four)	δ_3	-0.02			
		-0.2			
Transaction size	δ_4	0.00			
		-0.3			
Relative transaction size	δ_5	-0.10	-0.10 *		-0.07
		-1.3	*-1.6*		*-1.0*
Year of M&A	δ_6	0.00			
		-0.1			
Nobs	75				
Adj. R²		-0.03	0.02	0.01	0.00
F-stat. (p-value)		0.83	0.28	0.45	0.58
DW-statistic		2.20	2.20	2.25	2.17

***, **, and * indicate statistical significance at the 1%, 5%, and 10% levels, respectively. t-statistics in italic; CAB: cumulative abnormal beta shift.

While overall the regression models are statistically insignificant, we find weak evidence for a couple of potential drivers of abnormal beta changes. First, regression model M2 suggests a potential negative relationship between emerging market transactions and systematic risk. The negative coefficient is significant at the 10% level. Second, while we did not find an impact of the relative transaction size in the risk event study, regression model M2 though suggests a potential negative relationship between relative transaction size and abnormal beta change. Again, the negative coefficient is significant at the 10% level. Contrary to our expectations, we do not find any significant coefficients for cross-border transactions, transaction value, nor transactions by the "big four".

Our second multivariate regression analysis for the rivals takes the form outlined in Equation 5.5:

$$CAB_{rival} = \alpha + \delta_1 * MM + \delta_2 * TV + \delta_3 * RS + \delta_4 * Yr \quad (5.5)$$

Geographical scope: The results of the risk event study suggested a positive relationship between rival CAABs and M&A in mature markets. We include a respective dummy variable MM in our regression analysis in order to account for mature market transactions and to verify the preliminary risk analysis.

Transaction value and relative transaction size: While we did not find significant proof to suggest a relationship between acquirer CAABs and the size of acquirer or target, we include both transaction value (TV) and relative transaction size (RS) in our regression analysis.

Transaction date: The risk event study suggested a change in rival CAABs over time. We include the year of the transaction announcement, Yr, in the regression analysis in order to account for a potential dependency.

Table 5.9 presents the results of four regression models (M1-M4) on the rival CABs for the [0;5] event window. While, we find significant rival CAABs for several event windows we decide to again perform our second multivariate analysis with respect to the [0;5] event window to stay consistent with the previous analysis. The regression models differ with regard to the independent variables included. Models M1 to M3 are statistically significant at the 5% or 10% level (based on F-statistics). Explanatory power is higher than for the acquirer analysis with adjusted R-squared ranging between 2% and 7%. Autocorrelation issues can be ruled out as the Durbin-Watson-statistics are close to two in all cases. The constant, alpha, is insignificant for all three models.

Table 5.9: Multivariate Regressions on Rival CABs

Dependent variable		CAB [0;5]			
Model		M1	M2	M3	M4
constant	α	-23.6	-0.02	-0.02	-0.02
		-1.1	*-0.6*	*-0.6*	*-0.6*
Independent variables					
Target mature market	δ_1	0.06			0.11 *
		1.1			*1.7*
Transaction value	δ_2	0.00 ***	0.00 ***		
		-3.0	*-2.6*		
Relative transaction size	δ_3	0.34 ***	0.34 ***	0.20 *	
		3.5	*3.8*	*1.8*	
Year of M&A	δ_4	0.01			
		1.1			
Nobs	96				
adj. R²		0.06	0.07	0.04	0.02
F-stat. (p-value)		0.08 *	0.03 **	0.10 *	0.26
DW-statistic		1.74	1.71	1.67	1.75

***, **, and * indicate statistical significance at the 1%, 5%, and 10% levels, respectively. t-statistics in italic; *CAB*: cumulative abnormal beta shift.

The risk event study suggested a positive relationship between transactions in mature markets and rival *CAABs*. We do find evidence in the multivariate analysis to support such a relationship. However, evidence appears at low significance only testing the variable individually and the F-test does not approve the respective model M4 as generally significant. In addition, the multivariate analysis yields significant results with regard to the impact of transaction value and relative transaction size. The regression analysis suggests a negative impact of transaction value as indicated with negative t-statistics. Relative transaction size though shows a strongly positive effect on rival *CABs*. The corresponding coefficients are highly significant for two of three regression models. Finally, again contradicting the risk event study, the multivariate results do not suggest a change in *CABs* over time.

5.5. Conclusion

The objective of this study was to measure and analyze the abnormal short-term systematic risk shifts of acquirers and rivals in the brewing industry. For this purpose, a sample of 75 horizontal M&A transactions involving brewing companies between 1991 and 2010 was identified and analyzed using a combination of two approaches: the traditional event study methodology and the GARCH model based on Bollerslev (1986) and Engle (1982). Our results provide new insights into value implications of M&A transactions in the brewing industry through risk dynamics.

In order to provide a comprehensive review of short-term risk effects of M&A we analyzed a global dataset, which enabled us to specifically build and investigate relevant subsamples to determine key drivers of systematic risk. Given the dominance of the "big four" in the global brewing industry, we specifically analyzed abnormal risk as well as rival effects among them. In addition to the risk event study subsample analyses, the potential drivers were scrutinized by two multivariate regression models.

Our results indicate that on a total sample basis brewers do not experience significant abnormal risk shifts following the announcement of M&A transactions. These results are in line with long-term studies by Haugen and Langetieg (1975), Pettway and Yamada (1986), and Amihud, Dedong, and Saunders (2002) all of which provide evidence of no changes in systematic risk as a result of M&A announcements.

However, our results provide evidence for a number of drivers of systematic risk adjustments focusing on certain acquirer, target, or transaction characteristics within subsamples. Specifically, we find significantly negative acquirer abnormal risk shifts for cross-border transactions, in particular when targets are based in emerging market economies. These findings are in line with Kiymaz and Mukherjee (2001) and thus confirm the argument that diversification through cross-border transactions provides risk reduction and therefore value-enhancing opportunities. Moreover, we find evidence that suggests a negative relationship between systematic risk and transaction value.

Furthermore, we provide evidence that the "big four" experience significantly negative abnormal risk shifts when announcing an M&A transaction. At the same time, we find significantly positive abnormal risk shifts for close rivals missing out on a potential M&A opportunity, in particular when targets are based in mature markets where beer markets are generally fairly concentrated and thus M&A opportunities are limited. Moreover, our analysis suggests a positive relationship between rival companies' systematic risk and relative transaction size as well as a negative relationship with transaction value. Overall, our analysis on risk dynamics complements existing literature on short-term wealth effects, which generally assumes risk levels to be unaffected by M&A announcements and thus potentially misstates total shareholder value effects.

At the same time, it should be noted that there are also limitations to our study as our analysis is solely focused on the brewing industry. Due to the innovative and new risk event study methodology utilized there are no comparable results available. It could thus be questioned whether the determined results are truly industry driven or general M&A-related effects.

Having identified the negative short-term risk implications of cross-border M&A as well as determinant drivers, the question arises whether the short-term reductions in systematic risk are sustained in the long-run. So far there exists no empirical evidence analyzing the long-term implications of M&A on systematic risk of brewers. Hence, further empirical evidence should be provided by future research.

6. Conclusion

This doctoral thesis follows the research objective to improve our understanding of the capital market implications of mergers and acquisitions (M&A) in the global brewing industry. The detailed and thorough analysis of M&A in the sector becomes particularly interesting due to the strong consolidation process that has taken place in the brewing industry and dramatically reshaped its fundamental structures. M&A have and continue to be a major trend in the sector as multinational breweries seek to expand their activities into new emerging markets. At the same time, declining mature markets (in particular Western Europe) and resulting pressure on profit margins have encouraged brewers to engage in M&A, in order to gain in scale and benefit from synergies. In contrast to many other sectors, increased size provides brewers with a significant competitive advantage and hence it is not surprising that the global beer market today is dominated by large national/multinational brewers rather than local, regional brewers. The four largest brewers Anheuser-Busch Inbev, Heineken, SABMiller and Carlsberg (the "big four") control about 50% of the global beer market and compete fiercely for market share and potential acquisition opportunities.

Due to the dramatic changes in industry structure and the sector's unique competitive situation, this doctoral thesis analyzes the shareholder wealth and risk effects to acquiring brewers as well as rival companies missing out on acquisition opportunities. In order to achieve the research objective, this thesis provides a detailed industry analysis and conducts three individual empirical studies:

The first empirical study investigates the short-term wealth effects to acquiring brewers and close rival companies. Based on a sample of 69 takeover announcements between 1998 and 2010, significant positive announcement returns to acquiring brewers are found, which represents an outstanding attribute of the sector. At the same time, a number of determinant variables are detected that significantly impact short-term performance: the study finds a significant positive impact of domestic transactions as well as cross-border deals involving targets in emerging markets. Other identified value drivers include transaction size, acquirer size and the target's public status. Furthermore, the study provides evidence of significant negative rival effects across the "big four", when missing out on a potential M&A opportunity.

The second empirical study builds on the findings of the previous study and analyzes the long-term shareholder wealth effects of M&A on acquirers and rivals. The results show that in contrast to existing cross-industrial studies, which provide consistent evidence of value losses, acquirers in the brewing industry do not suffer from long-term value losses, following the engagement in M&A. Based on a sample of 66 transactions, the study provides evidence of abnormal performance that is insignificant if not significantly positive, thus suggesting that capital markets also value the successful search for acquisition targets in the long run. In addition, the study documents a significant negative impact of cross-border transactions and cash transactions on acquirer performance. On the other hand, no significant abnormal long-term returns are found for transactions by the "big four", neither to successful acquirers nor to rivals missing out. Overall, the results suggest that the determined short-term shareholder value effects are not sustainable in the long-run.

The third study shifts the focus of analysis on the risk implications of M&A. Specifically, the study sheds light on the question whether the M&A strategies employed by brewing companies are reflected by capital markets in the form of abnormal systematic risk shifts to acquiring brewers and rival companies. Based on a sample of 75 takeover announcements between 1998 and 2010, significant negative acquirer abnormal risk shifts are found for cross-border transactions, especially when targets are based in emerging markets. Furthermore, the study provides evidence of significant abnormal negative risk shifts to the "big four" when announcing an M&A transaction and significant positive abnormal risk shifts when missing out on a potential M&A transaction as rivals.

Overall, this doctoral thesis provides a comprehensive assessment of M&A in the brewing industry. I draw three major implications from my research, that are relevant to managers of brewing companies, investors and financial researchers:

First, the thesis identifies the brewing industry as an outlier industry with regards to shareholder wealth effects. In contrast to previous cross-industrial evidence, my study, across a number of different research approaches, shows that capital markets value the successful search for M&A opportunities in a consolidated global beer market, where size provides a material competitive advantage. The extraordinary synergy and efficiency potential is reflected by significant positive abnormal short-term performance of acquirers as well as non-negative long-term performance, both of which represent an outstanding attribute of the sector. Overall, these results imply that M&A in the sector are positively perceived by capital markets and thus provide a viable strategic option in the future.

Second, the thesis shows that the diversification strategies into international markets, as employed by many brewers, have a strong impact on the capital market risk profile of acquiring brewers. Cross-border transactions, in particular in emerging markets, result in significant short-term declines in acquirers' systematic risk. The risk-reducing effect implies that cross-border M&A in the sector present a feasible diversification option.

Third, the thesis empirically confirms the unique competitive situation among the leading brewing groups ("big four"). In a consolidated industry where it is becoming increasingly hard to find suitable M&A targets, capital markets punish the "big four", when missing out on an M&A opportunity to a major competitor. The effect of being a "transaction outsider" is reflected by significant negative short-term losses as well as significant increases in systematic risk. Overall, these results stress the importance of the "big four" to actively engage in M&A strategies.

In general, my findings seem very consistent with previous industry research (see e.g., Kerkvliet et al., 1998; Earlam et al., 2010; and Schwankl, 2008). Even though the thesis focuses solely on the brewing industry, the documented results underline the importance of industry-specific M&A analyses, as sector related potentials to generate value might differ among industries, leading to possibly biased finding of cross-industry examinations.

While the thesis addresses a number of important questions with regard to capital market effects of M&A in the brewing industry, the presented results give rise to new research issues. As the studies analyze the impact of M&A solely from a capital market perspective, future research could additionally examine whether the determined effects are also reflected in the operating performance of acquirers and rivals. In particular, case studies on individual M&A transactions by the "big four" could provide further insights with regard to post-merger acquirer and rival effects. Overall, given the expected continuation of the consolidation process in the sector, M&A in the brewing industry provide an interesting field for future research.

References

Adams, William J., 2006, Markets: Beer in Germany and the United States, Journal of Economic Perspectives 20/1, 189-205.

Agrawal, Anup, Jeffrey Jaffe, 2000, The Post-Merger Performance Puzzle, Advances in Mergers and Acquisitions, 7-41, Stamford: JAI Press.

Agrawal, Anup, Jeffrey Jaffe, Gershon Mandelker, 1992, The Post-Merger Performance of Acquiring Firms: A Re-Examination of an Anomaly, Journal of Finance 47/4, 1605-1621.

Akdogu, Evrim, 2009, Gaining a Competitive Edge Through Acquisitions: Evidence from the Telecommunications Industry, Journal of Corporate Finance 15/1, 99-112.

Amihud, Yakov, Gayle L. DeLong, Anthony Saunders, 2002, The Effects of Cross-border Bank Mergers on Bank Risk and Value, Journal of International Money and Finance 21/6, 857-877.

Andrade, Gregor, Mark, Mitchell, Erik, Stafford, 2001, New Evidence and Perspectives on Mergers, Journal of Economic Perspectives 15/2, 103-120.

Anheuser-Busch InBev, 2009, Annual Report, Brussels.

Asquith, Paul, Robert F. Bruner, David W. Mullins Jr., 1983, The Gains to Bidding Firms from Merger, Journal of Financial Economics 11/1-4, 121-139.

Aw, Michael, Robin Chatterjee, 2004, The Performance of UK Firms Acquiring Large Cross-border and Domestic Takeover Targets, Applied Financial Economics 14, 337-349.

Banal-Estañol, Albert, Marco Ottaviani, 2006, Mergers with Product Market Risk, Journal of Economics and Management Strategy 15/3, 577-608.

Banal-Estañol, Albert, Jo Seldeslachts, 2011, Merger Failures, Journal of Economics and Management Strategy 20/2, 589-624.

Barber Brad M., John D. Lyon, 1996, Detecting Abnormal Operating Performance: The Empirical Power and Specification of Test Statistics, Journal of Financial Economics 41/3, 359-399.

Bates, Theunis, 2009, A Lighter Brew: Nonalcoholic Beer, Time Website, Retrieved 1/19/12 from http://www.time.com/time/magazine/article/0,9171,1912354,00.html.

Beitel, Patrick, Dirk Schiereck, Mark Wahrenburg, 2004, Explaining M&A Success in European Banks, European Financial Management 10/1, 109-139.

Berry S. Keith, 2000, Excess Returns in Electric Utility Mergers During Transition to Competition, Journal of Regulatory Economics 18/2, 175-188.

Black, Ervin L., Thomas A. Carnes, Thomas Jandik, 2001, The Long-term Success of Cross-border Mergers and Acquisitions, Working Paper, University of Arkansas.

Boehmer, Ekkehart, Jim Musumeci, Annette B. Poulsen, 1991, Event-study Methodology under Conditions of Event-induced Variance, Journal of Financial Economics 30/2, 253-272.

Bollerslev, Tim, 1986, Generalized Autoregressive Conditional Heteroscedasticity, Journal of Econometrics 31/3, 307-327.

Bollerslev, Tim, 1990, Modelling the Coherence in Short-Run Nominal Exchange Rates: A Multivariate Generalized Arch Model, Review of Economics and Statistics 72/3, 498-505.

Bos, Theodore, Paul Newbold, 1984, An Empirical Investigation of the Possibility of Stochastic Systematic Risk in the Market Model, Journal of Business 57/1, 35-41.

Bradley, Michael, Anand Desai, E. Han Kim, 1988, Synergistic Gains from Corporate Acquisitions and their Division between the Stockholders of Target and Acquiring Firms, Journal of Financial Economics 21/1, 3-40.

Brown, Stephen J., Jerold B. Warner, 1985, Using daily stock returns: The case of event studies, Journal of Financial Economics 14/1, 3-31.

Brown, David, Michael Ryngaert, 1991, The Mode of Acquisition in Takeovers: Taxes and Asymmetric Information, Journal of Finance 46/2, 653-669.

Bruner, Robert F., 2002, Does M&A Pay? A Survey of Evidence for the Decision-Maker, Journal of Applied Finance 12/1, 48-69.

Brooks, Robert D., Robert W. Faff, Michael D. McKenzie, 1998, Time-Varying Beta Risk of Australian Industry Portfolios: A Comparison of Modeling Techniques, Australian Journal of Management 23/1, 1-22.

Carlsberg, 2009, Annual Report, Copenhagen.

Carroll, Glen R., Anand Swaminathan, 2000, Why the Microbrewery Movement? Organizational Dynamics of Resource Partitioning in the U.S. Brewing Industry, The American Journal of Sociology 106/3, 715-762.

Chatterjee, Sayan, Michael Lubatkin, 1990, Corporate mergers, stockholder diversification, and changes in systematic risk, Strategic Management Journal 11/4, 255-268.

Choi, Jongsoo, Jeffrey Russel, 2004, Economic Gains Around Mergers and Acquisitions in the Construction Industry of the United States of America, Canadian Journal of Civil Engineering 31/3, 513-525.

Clougherty, Joseph A., Tomaso Duso, 2009, The Impact of Horizontal Mergers on Rivals: Gains to Being Left Outside a Merger, Journal of Management Studies 46/8, 1365-1395.

Collins, Daniel W., Johannes Ledolter, Judy Rayburn, 1987, Some Further Evidence on the Stochastic Properties of Systematic Risk, Journal of Business 60/3, 425-448.

Conn, Robert L., 1985, A Re-examination of Merger Studies That Use the Capital Asset Pricing Model Methodology, Cambridge Journal of Economics 9/1, 43-56.

Conn, Robert C., Andy Cosh, Paul M. Guest, Alan Hughes, 2005, The Impact on UK Acquirers on Domestic, Cross-border, Public and Private Acquisitions, Journal of Business Finance & Accounting 32, 815-870.

Connell, Fred, Robert L. Conn, 1993, A Preliminary Analysis of Shifts in Market Model Regression Parameters in International Mergers Between US and British Firms: 1970-1980, Managerial Finance 19/1, 47-77.

Cowan, Arnold R., Anne M. Sergeant, 2001, Interacting Biases, Non-normal return Distributions and the Performance of Tests for Long-horizon Event Studies, Journal of Banking & Finance 25/4, 741-765.

Datamonitor, 2009, Marketwatch: Drinks – UK: Sales of Ales Increase, New York.

Datamonitor, 2011, Global Brewers – Industry Profile, New York.

Datta, Deepak K., George Puia, 1995, Cross-border Acquisitions: An Examination of the influence of Relatedness and Cultural Fit on Shareholder Value

Creation in U.S. Acquiring Firms, Management International Review 35/4, 337-359.

Davidson, III, Wallace N., Sharon H. Garrison, Glenn V. Henderson Jr., 1987, Examining Merger Synergy with the Capital Asset Pricing Model, Financial Review 22/2, 233-247.

DeRise, Jaon, Melissa Earlam, Renier Swanepoel, 2009, UBS Beer COGS Index: the trend continues (Equity Research Report), London.

Diageo, 2010, "Premiumisation", Diageo Website, Retrieved 1/19/2012, from http://www.diageo.com/Lists/Resources/Attachments/568/Premiumisation %20Seminar%20-%20FINAL%20PRINT.pdf.

Dodd, Peter, 1980, Merger Proposals, Management Discretion and Stockholder Wealth, Journal of Financial Economics 8/2, 105-137.

Earlam, Melissa, Jason DeRise, Renier Swanepoel, Kaumil Gajrawala and Poelma Zachary, 2010, European Beverages & Breweries - 2010: A flight to quality (Equity Research Report), London.

Ebneth, Oliver, Ludwig Theuvsen, 2007, Large Mergers and Acquisitions of European Brewing Groups – Event Study Evidence on Value Creation, Agri-business 23/3, 377-406.

Eckbo, B. Espen, 1983, Horizontal Mergers, Collusion, and Stockholder Wealth, Journal of Financial Economics 11/1-4, 241-273.

Eckbo, B. Espen, 1985, Mergers and the Market Concentration Doctrine: Evidence from the Capital Market, Journal of Business 58/3, 325-349.

Eckbo, B. Espen, Karin Thorburn, 2000, Gains to Bidder Firms Revisited: Domestic and Foreign Acquisitions in Canada, Journal of Financial & Quantitative Analysis 35/1, 1-25.

Elgers, Pieter T., John J. Clark, 1980, Merger Types and Shareholder Returns: Additional Evidence, Financial Management 9/2, 66-72.

Engle, Robert F., 1982, Autoregressive Conditional Heteroscedasticity with Estimates of the Variance of UK inflation, Econometrica 50/4, 987-1008.

Euromonitor International, 2009, Global Alcoholic Drinks: Beer - Opportunities in Niche Categories, London, United Kingdom.

Euromonitor International, 2010, Strategies for Growth in an Increasingly Conolidated Global Beer Market, London, United Kingdom.

Euromonitor International, 2011, Beer in Japan – Country Report, Euromonitor International Website, Retrieved 1/19/2012, from http://www.euro-monitor.com/beer-in-japan/report.

Fabozzi, Frank J., Jack C. Francis, 1978, Beta as a random coefficient, Journal of Financial and Quantitative Analysis 13/01, 101-116.

Faccio Mara, John J. Mcconnell, David Stolin, 2006, Returns to Acquirers of Listed and Unlisted Targets, Journal of Financial & Quantitative Analysis 41/1, 197-220.

Faff, Robert W., 2003, Creating Fama and French Factors with Style, Financial Review 38/2, 311-322.

Faff, Robert W., David Hillier, Joseph Hillier, 2000, Time Varying Beta Risk: An Analysis of Alternative Modelling Techniques, Journal of Business Finance and Accounting 27/5-6, 523-554.

Fama Eugene F., Kenneth R. French, 1992, The Cross-Section of Expected Stock Returns, The Journal of Finance 47/2, 427-465.

Fama, Eugen F., 1998, Market Efficiency, Long-Term Returns, and Behavioral Finance, Journal of Financial Economics 49/3, 283-306.

Fatemi, Ali M., 1984, Shareholder Benefits from Corporate International Diversification, Journal of Finance 39/5, 1325-1344.

Fee, Edward C., Shawn Thomas, 2004, Sources of Gains in Horizontal Mergers: Evidence from Customer, Supplier, and Rival Firms, Journal of Financial Economics 74/3, 423-460.

Ferris, Stephen P., Kwangwoo Park, 2002, How Different is the Long-Run Performance in the Telecommunications Industry? Advances in Financial Economics 7, 127-144.

Fletcher, Clementine, 2011, SABMiller's Mackay Predicts Further Beer Industry Consolidation, Retrieved 1/19/2012, from http://www.businessweek.com/news/2011-03-29/sabmiller-s-mackay-predicts-further-beer-industry-consolidation.html.

Funke, Christian, Timo Gebken, Lutz Johanning, Michel Gaston, 2008, Information Signaling and Competitive Effects of M&A: Long-Term Performance of Rival Companies, Working Paper, European Business School, Oestrich-Winkel.

Giannopoulos, Krzysztof, 1995, Estimating the Time-varying Components of International Stock Markets' Risk, European Journal of Finance 1/2, 129-164.

Gibbs, Mike, Matthew Webb, Komal Dhillon, 2010, European Beverages (Equity Research Report), London.

Goergen, Marc, Luc Renneboog, 2004, Shareholder Wealth Effects of European Domestic and Cross-border Takeover Bids, European Financial Management 10/1, 9-45.

Greenberg, Marc, Andrew Kieley, 2009, What's Facebook Drinking? (Equity Research Report), New York.

Greenberg, Marc, Jonathan Fell, Jose Yordan, 2010, Global Brewing Industry – Outlook & Consolidation Impliactions (Equity Research Report), New York.

Gregory, Alan, 1997, An Examination of the Long run Performance of UK Acquiring Firms, Journal of Business Finance & Accounting 24, 971-1002.

Haleblian, Jerayr, Sydney Finkelstein, 1999, The Influence of Organizational Acquisition Experience on Acquisition Performance: A Behavioral Learning Perspective, Administrative Science Quarterly 44/1, 29-56.

Hackbarth, Dirk, Erwan Morellec, 2008, Stock Returns in Mergers and Acquisitions, Journal of Finance 63/3, 1213-1252.

Harrington, Scott E., David G. Shrider, 2007, All Events Induce Variance: Analyzing Abnormal Returns When Effects Vary across Firms, Journal of Financial and Quantitative Analysis 42/1, 229-256.

Haugen, Robert A., Terence C. Langetieg, 1975, An Empirical Test for Synergism in Merger, Journal of Finance 30/4, 1003-1014.

Hawawini, Gabriel A., Itzhak Swary, 1991, Mergers and Acquisitions in the U.S. Banking Industry: Evidence from the Capital Markets, North-Holland, Amsterdam.

Heineken, 2009, Annual Report, Amsterdam.

Horowitz, Anna, Ira Horowitz, 1968, Entropy, Markov Processes and Competition in the Brewing Industry, Journal of Industrial Economics 16/3, 196-211.

Hull, John C., 2006, Options, Futures, and Other Derivatives, Sixth Edition, Pearson International, 461-478.

Iwasaki, Natsuko, Barry Seldon, Viktor Tremblay, 2008, Brewing Wars of Attrition for Profit (and Concentration), Review of Industrial Organization 33/4, 263-279.

Jaffe, Jeffrey, 1974, Special Information and Insider Trading, Journal of Business 47/3, 410-428.

Jarrell, Gregg A., Annette B. Poulsen, 1989, The Returns to Acquiring Firms in Tender Offers: Evidence from Three Decades, Financial Management 18/3, 12-19.

Jagannathan, Ravi, Zhenyu Wang, 1996, The Conditional CAPM and the Cross-Section of Expected Returns, Journal of Finance 51/1, 3-53.

Joehnk, Michael D., James F. Nielsen, 1974, The Effects of Conglomerate Merger Activity on Systematic Risk, Journal of Financial and Quantitative Analysis 9/2, 215-225.

Jones, David, 2010, Top Four Brewers Account for Over Half World's Beer, Reuters UK Web site, Retrieved 7/25/2011, 2011, from http://www.reuters.com/article/2010/02/08/beeridUSLDE61723K20100208.

Just-Drinks, 2007, Spotlight – Premium Beer Market set for Further Growth, Just-Drinks Website, Retrieved 1/19/12 from http://www.just-drinks.com/analysis/spotlight-premium-beer-market-set-for-further-growth_id91207.aspx.

Kerkvliet, Joe R., William Nebesky, Carol H. Tremblay, Viktor. J. Tremblay, 1998, Efficiency and Technological Change in the U.S. Brewing Industry, Journal of Productivity Analysis 10/3, 271-288.

Kiymaz, Halil, Tarun K. Mukherjee, 2001, Parameter Shifts When Measuring Wealth Effects in Cross-border Mergers, Global Finance Journal 12/2, 249-266.

Kothari, S. P., Jerold Warner, 2007, Econometrics of Event Studies, in: Eckbo, B.E. (Ed.), Handbook of Corporate Finance, Elsevier/North Holland, Amsterdam/Oxford.

Laabs, Jan-Peer, Dirk Schiereck, 2010, The Long-term Success of M&A in the Automotive Supply Industry: Determinants of Capital Market Performance, Journal of Economics and Finance 34/1, 61-88.

Langetieg, Terence C., Robert A. Haugen, Dean W. Wichern, 1980, Merger and Stockholder Risk, Journal of Financial and Quantitative Analysis 15/3, 689-717.

Lintner, John, 1965, The Valuation of Risky Assets and the Selection of Risky Investments in Stock Portfolios and Capital Budgets, Review of Economics and Statistics 47/1, 13-37.

Loughran, Tim, Jay Ritter, 2000, Uniformly Least Powerful Tests of Market Efficiency, Journal of Financial Economics 55/3, 361-389.

Loughran, Tim, Anand Vijh, 1997, Do Long-Term Shareholders Benefit From Corporate Acquisitions?, Journal of Finance 52/5, 1765-1790.

Lubatkin, Michael, Sayan Chatterjee, 1994, Extending Modern Portfolio Theory into the Domain of Corporate Diversification: Does it Apply?, Academy of Management Journal 37/1, 109-136.

Lubatkin, Michael, Hugh M. O'Neill, 1987, Merger Strategies and Capital Market risk, Academy of Management Journal 30/4, 665-684.

Lynk, William J., 1985, The Price and Output of Beer Revisited, Journal of Business 58/4, 433-437.

Lyon, John D., Brad M. Barber, Chih-Ling Tsai, 1999, Improved Methods for Tests of Long-Run Abnormal Stock Returns, Journal of Finance 54/1, 165-201.

MacKinlay, Craig A., 1997, Event Studies in Economics and Finance, Journal of Economic Literature 35/1, 13-39.

Mandelker, Gershon, 1974, Risk and return: The Case of Merging Firms, Journal of Financial Economics 1/4, 303-335.

Martynova, Marina, Luc Renneboog, 2006, Mergers and Acquisitions in Europe, Advances in Corporate Finance and Asset Pricing, 13–75, Amsterdam: Elsevier.

Mehta, Ramit, Dirk Schiereck, 2011, The Consolidation of the Brewing Industry and Wealth Implications from Mergers and Acquisitions, International Journal of Business and Finance Research, forthcoming.

Mentz, Markus, Dirk Schiereck, 2006, The Sources of Gains to International Takeovers: The Case of the Automotive Supply Industry, Working Paper, European Business School, Oestrich-Winkel.

Mintel, 2005, French and Germans Call Time on Drinking, Mintel Website, Retrieved 1/19/12 from http://www.mintel.com/press-centre/press-releases/50/french-and-germans-call-time-on-drinking.

Mintel, 2011, Little Cheer for Beer this Year as Revenue Sales fall £2.2 billion, Mintel Website, Retrieved 1/19/12 from http://www.mintel.com/press-centre/press-releases/795/little-cheer-for-beer-this-year-as-revenue-sales-fall-22-billion.

Mitchell, Mark, Erik Stafford, 2000, Managerial Decisions and Long-Term Stock Price Performance, Journal of Business 73, 287-329.

Moeller, Sara B., Frederik P. Schlingemann and René M. Stulz, 2003, Do Shareholders of Acquiring Firms Gain from Acquisitions?, NBER working paper No. 9523.

Monaghan, Gabrielle, Le-Min Lim, 2004, SABMiller Ends Harbin Bid - Bows Out to Anheuser-Busch, Bloomberg Website, Retrieved 1/19/12 from http://www.bloomberg.com/apps/news?pid=newsarchive&sid=alTxhB6e0t0s&refer=asia.

Morck, Randall, Bernard Yeung, 1992, Internalization: An Event Study Test, Journal of International Economics 33/1-2, 41-56.

Myers, Stewart C., Nicholas Majluf, 1984, Corporate Financing and Investment Decisions when Firms have Information that Investors Do Not Have, Journal Of Financial Economics 13/2, 187-221.

Peles, Yoram, 1971, Economies of Scale in Advertising Beer and Cigarettes, The Journal of Business 44/1, 32-37.

Pettway, Richard H., Takeshi Yamada, 1986, Mergers in Japan and Their Impacts upon Stockholders' Wealth, Financial Management 15/4, 43-52.

Porter, Michael E., 1980, Competitive Strategy, Free Press, New York.

Rabobank Group, 1999, Can Summer Warm up Beer and Wine Sales?, Rabobank Group Website, Retrieved 1/19/12 from http://www.rabobank.com/content/news/news_archive/063Cansummerwarmupbeerandwinesales.jsp.

Rau, Raghavendra, Theo Vermaelen, 1998, Glamour, Value and the Post-Acquisition Performance of Acquiring Firms, Journal of Financial Economics 49/2, 223-253.

Ritter, Jay, 1991, The Long-run Performance of Initial Public Offerings, Journal of Finance 46/1, 197-216.

SABMiller, 2009, Annual Report, London.

Schwankl, Matthias, 2008, M&A in der Deutschen Brauwirtschaft, BRAUINDUSTRIE, 12-15.

Schwert, William G., Paul J. Seguin, 1990, Heteroscedasticity in Stock Returns, Journal of Finance 45/4, 1129-1155.

Serra, Ana P, 2002, Event study tests: a brief survey, Working Paper, Faculdade de Economia da Universidade do Porto.

Shahrur, Husayn, 2005, Industry Structure and Horizontal takeovers: Analysis of wealth Effects on Rivals, Suppliers, and Corporate customers, Journal of Financial Economics 76/1, 61-98.

Sharpe, William F., 1964, Capital Asset Prices: A theory of Market Equilibrium Under Conditions of Risk, Journal of Finance 19/3, 425-442.

Slade, Margaret, 2004, Market Power and Joint Dominance in U.K. Brewing, Journal of Industrial Economics 52/1, 133-163.

Shepherd, John, 1994, Brewers Will Have to Cut Costs: Fierce Competition Continues as National Consumption Ceclines, The Independent Website, Retrieved 1/19/12 from http://www.independent.co.uk/news/business/brew-ers-will-have-to-cut-costs-fierce-competition-continues-as-national-consumption-declines-1394146.html.

Snyder, Christopher M., 1996, A Dynamic Theory of Countervailing Power, RAND Journal of Economics 27/4, 747-769.

Song, Moon H., Ralph A. Walkling, 2000, Abnormal Returns to Rivals of Acquisition Targets: A test of the "Acquisition Probability Hypothesis", Journal of Financial Economics 55/2, 143-171.

Spain, William, 2008, Will Sports Lose one of its Biggest Boosters?, Wall Street Journal – Market Watch Website, Retrieved 1/19/12 from http://www.marketwatch.com/story/inbev-takeover-spotlights-anheuser-buschs-big-ad-budget.

Stout, Jean C., 2007, Big Brewers Pour Into Emerging Markets, Business Week Website, Retrieved 1/19/2012, from http://www.businessweek.com/investor/content/jun2007/pi20070604_644419.htm.

Sunder, Shyam, 1980, Stationarity of Market Risk: Random Coefficients Test for Individual Stocks, Journal of Finance 35/4, 883-896.

Tholl, Gregor, 1999, Der Siegeszug des Weizenbiers, N24 Website, Retrieved 1/19/2012, from http://www.n24.de/news/newsitem_5412060.html.

Tremblay, Viktor J., Carol H. Tremblay, 1988, The Determinants of Horizontal Acquisitions: Evidence from the US Brewing Industry, Journal of Industrial Economics 37/1, 21-45.

Tremblay, Viktor J., Carol H. Tremblay, 2009, The U.S. Brewing Industry - Data and Economic Analysis. Cambridge.

Wansley, James W., William R. Lane, Ho C. Yang, 1983, Abnormal Returns to Acquired Firms By Type of Acquisition and Method of Payment, Financial Management 12/3, 16-22.

Williams, Jonathan, Angel Liao, 2008, The Search for Value: Cross-border Bank M&A in Emerging Markets, Comparative Economic Studies, 50/2, 274-296.

World Health Organization, 2011, A Summary of Global Status Report on Alcohol, World Health Organization Website, Retrieved 1/19/2012, from http://www.who.int/substance_abuse/publications/en/globalstatussummary.pdf.

Zhu, Peng C., Shavin Malhotra, 2008, Announcement Effect and Price Pressure: An Empirical Study of Cross-Border Acquisitions by Indian Firms, Journal of Finance and Economics 13, 24-41.

Appendix

Table A4.1: BHARs of Cross-border Deals: Mature vs. Emerging Market Transactions

	Holding Period	6 Months	12 Months	18 Months	24 Months
Mature Market	BHAR	-6.82%	-8.34%	-12.30%	-14.45%
	t-value	-1.51	-1.64	-1.53	-1.55
	p-value	0.14	0.12	0.14	0.14
	N	22	22	22	21
Emerging Market	BHAR	0.17%	3.12%	0.33%	-11.47%
	t-value	0.05	0.53	0.04	-1.17
	p-value	0.96	0.60	0.97	0.25
	N	30	30	29	18
Difference	Delta	-6.99%	-11.46%	-12.63%	-2.98%
	t-value	-1.19	-1.47	-1.15	-0.22
	p-value	0.24	0.15	0.26	0.83

This table shows the average Buy-and-Hold Abormal Returns to acquiring companies in cross-border mergers and acquisitions in the brewing industry differentiated by mature market and emerging market transactions. Abnormal returns are derived using a control-firm matching approach as proposed by Lyon, Barber, and Tsai (1999). All acquirers between 1998-2010 are included for which the relevant matching information is available (market values and market-to-book ratios). Statistical significance at the 10%, 5%, and 1% level is denoted by *, **, *** respectively, and is tested using a standard t-test. Statistical significance of mean differences is tested using the two groups difference of means test as proposed by Cowan and Sergeant (2001).

Table A4.2: BHARs: Public vs. Private Targets

	Holding Period	6 Months	12 Months	18 Months	24 Months
Public	BHAR	-2.59%	-0.90%	-3.87%	-10.83%
	t-value	-0.51	-0.15	-0.54	-1.29
	p-value	0.61	0.88	0.59	0.21
	N	28	28	27	27
Private	BHAR	-1.73%	2.84%	0.21%	0.15%
	t-value	-0.62	0.58	0.03	0.02
	p-value	0.54	0.57	0.98	0.99
	N	38	38	38	35
Difference	Delta	-0.86%	-3.74%	-4.07%	-10.98%
	t-value	-0.15	-0.48	-0.40	-0.90
	p-value	0.88	0.63	0.69	0.37

This table shows the average Buy-and-Hold Abormal Returns to acquiring companies in mergers and acquisitions in the brewing industry differentiated by the target's public status. Abnormal returns are derived using a control-firm matching approach as proposed by Lyon, Barber, and Tsai (1999). All acquirers between 1998-2010 are included for which the relevant matching information is available (market values and market-to-book ratios). Statistical significance at the 10%, 5%, and 1% level is denoted by *, **, *** respectively, and is tested using a standard t-test. Statistical significance of mean differences is tested using the two groups difference of means test as proposed by Cowan and Sergeant (2001).

Table A4.3: BHARs: "big four" vs. Other Acquirers

	Holding Period	6 Months	12 Months	18 Months	24 Months
"big four"	BHAR	-3.90%	-3.27%	-4.06%	-11.65%
	t-value	-1.05	-0.77	-0.70	-1.28
	p-value	0.30	0.45	0.49	0.21
	N	35	35	34	34
Other Acquirers	BHAR	-0.06%	6.36%	1.33%	3.89%
	t-value	-0.02	0.99	0.15	0.49
	p-value	0.99	0.33	0.88	0.63
	N	31	31	31	28
Difference	Delta	-3.83%	-9.62%	-5.39%	-15.55%
	t-value	-0.72	-1.25	-0.51	-1.29
	p-value	0.47	0.22	0.61	0.20

This table compares the average Buy-and-Hold Abormal Returns to the "big four" (Anheuser-Busch Inbev, SABMiller, Heineken and Carlsberg) with all other acquirers in the sample. Abnormal returns are derived using a control-firm matching approach as proposed by Lyon, Barber, and Tsai (1999). All acquirers between 1998-2010 are included for which the relevant matching information is available (market values and market-to-book ratios). Statistical significance at the 10%, 5%, and 1% level is denoted by *, **, *** respectively, and is tested using a standard t-test. Statistical significance of mean differences is tested using the two groups difference of means test as proposed by Cowan and Sergeant (2001).

Table A4.4: Calendar Time Portfolio of "big four"

	Holding Period	6 Months		12 Months		18 Months		24 Months	
"big four"	Alpha (t-value)	-1.18% (-2.13)	**	-0.19% (-0.37)		-0.38% (-0.78)		-0.26% (-0.55)	
	Beta (t-value)	0.60 (6.17)	***	0.79 (8.64)	***	0.74 (8.52)	***	0.74 (8.81)	***
	SMB (t-value)	0.15 (0.65)		0.17 (0.81)		0.24 (1.19)		0.26 (-0.55)	
	HML (t-value)	0.53 (2.85)	***	0.68 (3.91)	***	0.52 (3.12)	***	0.57 (3.55)	***
	F-value	13.54	***	26.39	***	25.97	***	27.99	***
	R^2	0.22		0.35		0.34		0.36	
Rivals	Alpha (t-value)	-1.62% (-3.73)	***	-0.43% (-0.84)		-0.49% (-0.96)		-0.46% (-0.93)	
	Beta (t-value)	0.51 (6.64)	***	0.89 (9.72)	***	0.82 (9.21)	***	0.86 (9.73)	***
	SMB (t-value)	0.51 (2.87)	***	0.82 (3.85)	***	0.79 (3.78)	***	0.75 (3.63)	***
	HML (t-value)	0.35 (2.42)	**	0.64 (3.71)	***	0.66 (3.90)	***	0.67 (3.97)	***
	F-value	19.44	***	40.38	***	36.62	***	39.65	***
	R^2	0.28		0.45		0.43		0.45	

This table shows the calendar time portfolio returns, based on the Fama-French-3-Factor-model, for acquisitions by the "big four" (Anheuser-Busch Inbev, SABMiller, Heineken and Carlsberg). Additionally, the table reports the calendar time portfolio returns to the "big four" when missing out as rivals on potential acquisition opportunities. Statistical significance at the 10%, 5%, and 1% level is denoted by *, **, *** respectively, and is tested using a standard t-test. Furthermore, statistical quality of the model is indicated by the determination coefficient R^2 and the corresponding f-statistic

Table A4.5: Multivariate Regression on Acquirer BHAR

Dependent variable: Acquirer BHAR

Holding Period in months		6m		12m		18m		24m
Constant	α	-0.05		0.10		0.04		0.26
		-0.73		*1.08*		*0.31*		*1.57*
Independent variables								
Cross-border M&A	γ_1	-0.09		-0.21	*	-0.21		-0.35
		-1.13		*-1.94*		*-1.43*		*-1.84*
Target emerging market	γ_2	0.06		0.02		0.01		-0.13
		1.14		*0.23*		*0.11*		*-0.93*
Transaction value	γ_3	0.00		0.00		0.00		-0.00
		1.28		*1.10*		*0.24*		*-0.35*
Public target	γ_4	-0.04		-0.11		-0.12		-0.18
		-0.65		*-1.42*		*-1.10*		*-1.34*
"big four"	γ_5	0.10		-0.03		0.12		0.15
		1.22		*-0.24*		*0.77*		*0.79*
Share component	γ_6	0.16	**	*0.18*	**	0.37	***	0.16
		2.60		*2.00*		*2.96*		*1.01*
1998 – 2003	γ_7	0.06		0.04		-0.07		-0.03
		1.01		*0.53*		*-0.59*		*-0.18*
R^2 (adj.)		0.13		0.09		0.09		0.06
F-stat. (p-value)		0.03	**	0.09	*	0.08	*	0.18
DW-statistic		1.82		1.69		1.53		1.89

This table shows the results of cross-sectional regressions of acquirer buy-and-hold abnormal returns (BHAR) on a host of explanatory variables. ***, **, and * indicate statistical significance at the 1%, 5%, and 10% levels, respectively. t- statistics in italic; *BHAR*: buy-and-hold abnormal return

Table A5.1: Acquirer CAABs of Overall Transaction Sample

CAAB [-x; +y]	Mean		Median	KS-test	p	Wx-test	p	t-test	p	N
0 0	-0.01		0.00	0.12	0.23	-1.03	0.30	-0.02	0.32	75
1 1	-0.02		0.00	0.10	0.36	-1.24	0.22	-0.06	0.22	75
2 2	-0.04		-0.01	0.10	0.39	-1.29	0.20	-0.10	0.18	75
5 5	-0.09		-0.03	0.10	0.44	-1.39	0.16	-0.23	0.18	75
10 10	-0.15		-0.02	0.09	0.51	-1.30	0.19	-0.40	0.23	75
20 20	-0.24		-0.13	0.10	0.38	-1.29	0.20	-0.68	0.29	75
0 1	-0.02		0.00	0.13	0.13	-1.25	0.21	-0.04	0.16	75
0 2	-0.03		-0.01	0.13	0.17	-1.48	0.14	-0.07	0.11	75
0 5	-0.07	*	-0.03	0.12	0.21	-1.61	0.11	-0.16	0.10	75
0 10	-0.12		-0.08	0.10	0.47	-1.44	0.15	-0.29	0.15	75
0 20	-0.18		-0.05	0.08	0.64	-1.19	0.23	-0.52	0.28	75

***, **, and * indicate statistical significance at the 1%, 5%, and 10% levels, respectively. *CAAB*: cumulative average abnormal beta shift, KS test: Kolmogorov-Smirnov test - p-value ≤ 0.05 indicates normal distribution at the 5% significance level, Wx-test: Wilcoxon signed rank test as non-parametric median test, t-test: std. parametric mean test.

Table A5.2: Acquirer CAABs of Domestic Transactions

CAAB [-x; +y]	Mean	Median	KS-test	p	Wx-test	p	t-test	p	N
0 0	0.00	0.00	0.23	0.15	-0.46	0.65	-0.03	0.81	24
1 1	-0.01	0.01	0.23	0.14	-0.69	0.49	-0.08	0.86	24
2 2	-0.01	0.01	0.24	0.11	-0.49	0.63	-0.13	0.86	24
5 5	-0.02	0.02	0.24	0.11	-0.57	0.57	-0.28	0.86	24
10 10	0.02	0.10	0.19	0.29	-0.89	0.38	-0.46	0.94	24
20 20	0.16	0.11	0.15	0.57	-0.89	0.38	-0.68	0.71	24
0 1	-0.01	0.00	0.24	0.11	-0.69	0.49	-0.06	0.77	24
0 2	-0.01	0.01	0.23	0.13	-0.34	0.73	-0.09	0.75	24
0 5	-0.02	0.01	0.21	0.22	-0.20	0.84	-0.18	0.78	24
0 10	0.01	0.02	0.16	0.55	-0.57	0.57	-0.28	0.92	24
0 20	0.16	0.04	0.16	0.55	-0.80	0.42	-0.43	0.59	24

***, **, and * indicate statistical significance at the 1%, 5%, and 10% levels, respectively. *CAAB*: cumulative average abnormal beta shift, KS test: Kolmogorov-Smirnov test - p-value ≤ 0.05 indicates normal distribution at the 5% significance level, Wx-test: Wilcoxon signed rank test as non-parametric median test, t-test: std. parametric mean test.

Table A5.3: Acquirer CAABs by Relative Size of Transaction

CAAB	[-x; +y]		Mean	Median		KS-test	p	Wx-test	p	t-test	p	n
Bottom 30 (<6%)	0	0	-0.01	-0.01	*	0.17	0.33	-1.68	0.09	-0.03	0.26	30
	1	1	-0.03	-0.03		0.15	0.44	-1.53	0.13	-0.09	0.29	30
	2	2	-0.05	-0.05		0.15	0.48	-1.51	0.13	-0.15	0.28	30
	5	5	-0.12	-0.11		0.14	0.55	-1.57	0.12	-0.33	0.22	30
	10	10	-0.23	-0.22		0.15	0.45	-1.57	0.12	-0.61	0.21	30
	20	20	-0.44	-0.51	*	0.12	0.74	-1.64	0.10	-1.06	0.16	30
	0	1	-0.02	-0.02		0.16	0.41	-1.55	0.12	-0.06	0.25	30
	0	2	-0.04	-0.03	*	0.14	0.57	-1.66	0.10	-0.10	0.22	30
	0	5	-0.08	-0.07	*	0.15	0.50	-1.82	0.07	-0.20	0.16	30
	0	10	-0.16	-0.14	*	0.12	0.69	-1.66	0.10	-0.38	0.13	30
	0	20	-0.32	-0.28	*	0.10	0.92	-1.68	0.09	-0.74	0.13	30
Top 30 (19%<430%)	0	0	0.00	0.00		0.11	0.85	-0.20	0.85	-0.01	0.94	30
	1	1	-0.02	0.00		0.12	0.74	-0.42	0.67	-0.07	0.51	30
	2	2	-0.04	-0.01		0.13	0.65	-0.59	0.56	-0.12	0.39	30
	5	5	-0.09	-0.04		0.13	0.69	-0.73	0.47	-0.30	0.35	30
	10	10	-0.18	-0.10		0.13	0.63	-0.75	0.45	-0.56	0.36	30
	20	20	-0.26	-0.17		0.12	0.70	-0.65	0.52	-0.96	0.46	30
	0	1	-0.02	0.00		0.15	0.50	-0.36	0.72	-0.05	0.41	30
	0	2	-0.03	-0.01		0.16	0.37	-0.71	0.48	-0.10	0.30	30
	0	5	-0.09	-0.02		0.16	0.42	-0.89	0.37	-0.24	0.25	30
	0	10	-0.15	-0.07		0.15	0.50	-0.89	0.37	-0.45	0.30	30
	0	20	-0.21	-0.04		0.15	0.48	-0.59	0.56	-0.80	0.47	30

***, **, and * indicate statistical significance at the 1%, 5%, and 10% levels, respectively. *CAAB*: cumulative average abnormal beta shift, KS test: Kolmogorov-Smirnov test - p-value ≤ 0.05 indicates normal distribution at the 5% significance level, Wx-test: Wilcoxon signed rank test as non-parametric median test, t-test: std. parametric mean test.

Table A5.4: Acquirer CAABs by Date of Transaction Announcement

CAAB	[-x; +y]	Mean	Median	KS-test	p	Wx-test	p	t-test	p	n
1991 - 2003	0 0	-0.01	0.00	0.12	0.64	-0.88	0.38	-0.02	0.31	36
	1 1	-0.02	-0.01	0.10	0.82	-0.86	0.39	-0.07	0.34	36
	2 2	-0.04	-0.01	0.09	0.89	-0.90	0.37	-0.12	0.32	36
	5 5	-0.08	-0.03	0.11	0.78	-1.04	0.30	-0.25	0.35	36
	10 10	-0.12	-0.01	0.12	0.64	-0.88	0.38	-0.44	0.45	36
	20 20	-0.18	0.00	0.17	0.22	-0.79	0.43	-0.76	0.54	36
	0 1	-0.02	0.00	0.11	0.74	-0.93	0.35	-0.05	0.27	36
	0 2	-0.03	-0.01	0.10	0.80	-1.02	0.31	-0.08	0.25	36
	0 5	-0.06	-0.02	0.12	0.67	-1.27	0.20	-0.16	0.26	36
	0 10	-0.08	-0.04	0.12	0.66	-1.05	0.29	-0.29	0.42	36
	0 20	-0.12	-0.03	0.15	0.34	-0.64	0.52	-0.54	0.58	36
2004 - 2010	0 0	0.00	0.00	0.16	0.27	-0.64	0.52	-0.02	0.65	39
	1 1	-0.02	0.00	0.13	0.46	-0.92	0.36	-0.08	0.42	39
	2 2	-0.04	0.00	0.12	0.54	-0.89	0.37	-0.14	0.36	39
	5 5	-0.10	-0.02	0.12	0.54	-0.93	0.35	-0.32	0.33	39
	10 10	-0.18	-0.15	0.11	0.72	-0.91	0.36	-0.57	0.36	39
	20 20	-0.29	-0.47	0.12	0.64	-0.92	0.36	-0.99	0.40	39
	0 1	-0.02	0.00	0.15	0.28	-0.84	0.40	-0.06	0.35	39
	0 2	-0.04	-0.01	0.15	0.29	-1.00	0.32	-0.10	0.27	39
	0 5	-0.09	-0.06	0.15	0.33	-1.12	0.26	-0.23	0.22	39
	0 10	-0.16	-0.12	0.13	0.49	-0.98	0.33	-0.43	0.24	39
	0 20	-0.25	-0.27	0.09	0.86	-0.84	0.40	-0.78	0.36	39

***, **, and * indicate statistical significance at the 1%, 5%, and 10% levels, respectively. *CAAB*: cumulative average abnormal beta shift, KS test: Kolmogorov-Smirnov test - p-value ≤ 0.05 indicates normal distribution at the 5% significance level, Wx-test: Wilcoxon signed rank test as non-parametric median test, t-test: std. parametric mean test.

Table A5.5: Rival CAABs by Transaction Value

CAAB	[-x; +y]	Mean	Median		KS-test	p	Wx-test	p	t-test	p	n
Bottom 30 (50M≤132M)	0 0	0.01	0.01	**	0.14	0.55	-2.01	0.04	0.00	0.21	30
	1 1	0.02	0.02	*	0.13	0.69	-1.64	0.10	-0.02	0.34	30
	2 2	0.02	0.03		0.16	0.38	-1.61	0.11	-0.04	0.51	30
	5 5	0.03	0.05		0.19	0.18	-1.47	0.14	-0.12	0.67	30
	10 10	0.06	0.09		0.17	0.31	-1.20	0.23	-0.22	0.66	30
	20 20	0.11	0.15		0.18	0.26	-1.18	0.24	-0.39	0.66	30
	0 1	0.01	0.02	*	0.14	0.56	-1.70	0.09	-0.01	0.32	30
	0 2	0.01	0.02		0.17	0.29	-1.51	0.13	-0.03	0.54	30
	0 5	0.02	0.04		0.16	0.37	-1.31	0.19	-0.07	0.56	30
	0 10	0.05	0.05		0.14	0.52	-1.31	0.19	-0.13	0.56	30
	0 20	0.08	0.06		0.19	0.22	-1.20	0.23	-0.32	0.69	30
Top 30 (646M≤53B)	0 0	0.01	0.01		0.11	0.79	-1.35	0.18	-0.01	0.20	30
	1 1	0.04	0.03		0.10	0.89	-1.39	0.17	-0.02	0.19	30
	2 2	0.07	0.04		0.11	0.84	-1.45	0.15	-0.03	0.18	30
	5 5	0.14	0.07		0.11	0.83	-1.39	0.17	-0.07	0.19	30
	10 10	0.22	0.16		0.13	0.66	-1.33	0.18	-0.16	0.25	30
	20 20	0.33	0.39		0.14	0.58	-1.12	0.26	-0.34	0.32	30
	0 1	0.03	0.02		0.11	0.84	-1.49	0.14	-0.01	0.16	30
	0 2	0.04	0.03		0.12	0.76	-1.47	0.14	-0.02	0.17	30
	0 5	0.08	0.07		0.10	0.90	-1.24	0.21	-0.05	0.21	30
	0 10	0.10	0.18		0.11	0.85	-1.02	0.31	-0.14	0.41	30
	0 20	0.11	0.32		0.14	0.57	-0.87	0.38	-0.36	0.63	30

***, **, and * indicate statistical significance at the 1%, 5%, and 10% levels, respectively. *CAAB*: cumulative average abnormal beta shift, KS test: Kolmogorov-Smirnov test - p-value ≤ 0.05 indicates normal distribution at the 5% significance level, Wx-test: Wilcoxon signed rank test as non-parametric median test, t-test: std. parametric mean test.

Table A5.6: Rival CAABs by Date of Transaction Announcement

CAAB	[-x; +y]	Mean	Median		KS-test	p	Wx-test	p	t-test	p	n
1991 - 2003	0　0	0.01	0.01	**	0.15	0.34	-2.04	0.04	0.00	0.06	37
	1　1	0.04	0.02	**	0.15	0.36	-2.01	0.04	0.00	0.06	37
	2　2	0.06	0.03	*	0.12	0.66	-1.88	0.06	0.00	0.05	37
	5　5	0.14	0.06	*	0.12	0.65	-1.70	0.09	-0.01	0.06	37
	10　10	0.25	0.09	*	0.13	0.53	-1.70	0.09	-0.01	0.05	37
	20　20	0.52	0.35	**	0.11	0.73	-2.07	0.04	0.06	0.03	37
	0　1	0.03	0.01	**	0.14	0.41	-2.23	0.03	0.00	0.05	37
	0　2	0.04	0.02	**	0.13	0.53	-1.98	0.05	0.00	0.06	37
	0　5	0.07	0.04	*	0.09	0.86	-1.68	0.09	-0.01	0.10	37
	0　10	0.12	0.05		0.08	0.94	-1.50	0.13	-0.03	0.12	37
	0　20	0.28	0.18		0.08	0.94	-1.59	0.11	-0.04	0.08	37
2004 - 2010	0　0	0.00	0.01		0.20	0.02	-1.09	0.28	-0.02	0.79	59
	1　1	-0.01	0.02		0.19	0.02	-0.95	0.34	-0.05	0.73	59
	2　2	-0.02	0.03		0.18	0.04	-0.91	0.37	-0.09	0.66	59
	5　5	-0.04	0.05		0.20	0.02	-0.82	0.41	-0.20	0.58	59
	10　10	-0.10	0.09		0.20	0.02	-0.60	0.55	-0.40	0.49	59
	20　20	-0.22	0.12		0.20	0.02	-0.32	0.75	-0.75	0.42	59
	0　1	0.00	0.01		0.19	0.03	-0.93	0.35	-0.03	0.77	59
	0　2	-0.01	0.02		0.17	0.06	-0.84	0.40	-0.05	0.71	59
	0　5	-0.02	0.04		0.16	0.07	-0.66	0.51	-0.11	0.66	59
	0　10	-0.05	0.04		0.14	0.15	-0.40	0.69	-0.22	0.56	59
	0　20	-0.12	0.03		0.17	0.05	-0.07	0.95	-0.45	0.45	59

***, **, and * indicate statistical significance at the 1%, 5%, and 10% levels, respectively. *CAAB*: cumulative average abnormal beta shift, KS test: Kolmogorov-Smirnov test - p-value ≤ 0.05 indicates normal distribution at the 5% significance level, Wx-test: Wilcoxon signed rank test as non-parametric median test, t-test: std. parametric mean test.

Corporate Finance and Governance

Herausgegeben von Dirk Schiereck

www.peterlang.de

Peter Lang · Internationaler Verlag der Wissenschaften

Luciano Segreto / Hubert Bonin / Andrzej K. Kozminski /
Carles Manera / Manfred Pohl (eds.)

European Business and Brand Building

Bruxelles, Bern, Berlin, Frankfurt am Main, New York, Oxford, Wien, 2012.
264 pp. 19 ill.

ISBN 978-90-5201-793-8 · pb. € 37,30*
ISBN 978-3-0352-6142-4 (eBook) · € 41,53*

A strong brand is a key factor in business success, both in the short-term
and in the long-term. Brands help to provide a better understanding of the
corporate and commercial culture of different firms. A brand reveals the
knowledge capital held by a company, but also often reflects the perception
of the firm held by consumers and stake-holders. The book explores the
historical process of building some of the most famous brands among
European businesses and examines the extent to which the brands have
contributed to the image of the firms and their differentiation against
competitors in the industry.

Contents: Andrzej K. Kozminski: Preface · Manfred Pohl: Foreword · Luciano
Segreto: Caution: Brands at Work! Branding between Time, History,
and Financial Markets · Peter Miskell: Unilever and Its Brands since the
1950s. Competitive Threats and Strategic Responses · Dominique Barjot/
Francesca Tesi: The Building of Michelin's Corporate Image and Brand ·
Laurent Tissot: Suchard. A Swiss Chocolate Brand Somewhere in between
Traditional and Modern Cultures · Claire Desbois-Thibault: Champagne. A
Distinguished Wine? The Creation of Champagne's Brand Image · Hubert
Bonin: A Reassessment of the Business History of the French Luxury Sector.
The Emergence of a New Business Model and a Renewed Corporate Image
(from the 1970s) · Elisabetta Merlo: The Ascendance of the Italian Fashion
Brands (1970-2000) · Carles Manera/Jaume Garau-Taberner: The Invention of
the Camper Brand. Brand Building of Mallorca Shoe-Manufacturing · Xoán
Carmona Badía: Corporate Growth and Changes in Brand Identity. The Case
of the Zara Group · and more

*The e-price includes German tax rate. Prices are subject to change without notice

Frankfurt am Main · Berlin · Bern · Bruxelles · New York · Oxford · Wien
Distribution: Verlag Peter Lang AG
Moosstr. 1, CH-2542 Pieterlen
Telefax 0041 (0)32/3761727
E-Mail info@peterlang.com

40 Years of Academic Publishing
Homepage http://www.peterlang.com